石油和化工行业"十四五"规划教材

生物专业英语

Professional English for Biology

第二版

姜巨全　主编

化学工业出版社

·北京·

内容简介

《生物专业英语》第一版自2011年8月出版以来,被多所高等院校用作生物专业英语课程教材,广受好评。新版入选石油和化工行业"十四五"规划教材,根据用书高校学生反馈建议并针对生物学领域最新发展趋势,对部分内容进行小幅修订,同时在化工教育网上增加了电子课件。新版遵循第一版的原则和整体构架,具有以下特点:

(1) 专业英语内容选材丰富、知识点较新,各章节内容比例适当、难易程度适中;
(2) 在讲解专业英语词汇时注重讲解它们的准确含义及其具体的使用语境;
(3) 补充阅读材料的选取多以各学科最新的研究进展或报道为主;
(4) 每章起始设置中文导读,每节设置重点疑难句子的中文翻译;
(5) 课后习题多样化。

本教材适合高等院校生物科学、生物工程、生物技术等各个专业本科及研究生用于专业英语学习。

图书在版编目(CIP)数据

生物专业英语/姜巨全主编. —2版. —北京:化学工业出版社,2024.4
石油和化工行业"十四五"规划教材
ISBN 978-7-122-45293-1

Ⅰ.①生… Ⅱ.①姜… Ⅲ.①生物学-英语-高等学校-教材 Ⅳ.①Q

中国国家版本馆 CIP 数据核字(2024)第 059461 号

责任编辑:傅四周 文字编辑:刘洋洋
责任校对:王鹏飞 装帧设计:王晓宇

出版发行:化学工业出版社
　　　　(北京市东城区青年湖南街13号　邮政编码100011)
印　　刷:北京云浩印刷有限责任公司
装　　订:三河市振勇印装有限公司
787mm×1092mm　1/16　印张 11¼　字数 266 千字
2024年7月北京第2版第1次印刷

购书咨询:010-64518888　　　售后服务:010-64518899
网　　址:http://www.cip.com.cn
凡购买本书,如有缺损质量问题,本社销售中心负责调换。

定　价:45.00元　　　　　　　　版权所有　违者必究

作者名单

主　编　姜巨全

副主编　李春红　胡小梅　王洪军　孟　琳　邵　丽　王　浩

主　审　胡宝忠

编　者（按工作量排名）

　　　　姜巨全（东北农业大学）

　　　　李春红（哈尔滨医科大学附属肿瘤医院内科）

　　　　胡小梅（东北农业大学）

　　　　王洪军（哈尔滨医科大学附属第二医院神经外科）

　　　　孟　琳（东北农业大学）

　　　　邵　丽（东北农业大学）

　　　　王　浩（东北农业大学）

　　　　谷春涛（东北农业大学）

　　　　王艳红（黑龙江八一农垦大学）

　　　　张　瑞（中国水产科学研究院黑龙江水产研究所）

　　　　陈慧文（东北农业大学）

　　　　董士嘉（哈尔滨学院）

　　　　王多佳（东北农业大学）

前　言

　　《生物专业英语》（第二版）是石油和化工行业"十四五"规划教材，主编姜巨全教授现为首届国家"神农英才"计划——神农青年英才、黑龙江省首届"青年龙江学者"特聘教授。姜巨全教授在美国著名私立大学维克森林大学（Wake Forest University）以博士后身份学习工作近五年时间，主要从事分子生物学有关的科研工作，因此，具有丰富的普通英语和生物专业英语学习以及使用经验，对如何讲授生物专业英语以及讲授哪些专业英语知识具有很深的体会。自2009年回国被聘为东北农业大学教授以来，姜巨全教授一直从事生命科学学院生物专业英语的教学工作，并在《自然通讯》（Nature Communications）等国际知名期刊以第一/通讯作者发表30余篇SCI论文，非常了解国内生物学专业学生在本科专业英语知识学习以及研究生培养阶段专业英语学习方面所面临的问题。

　　生物专业英语普遍作为生物学及相关专业的专业必修课，生物专业英语的学习对生物学本科生及研究生至关重要，在学生从普通英语学习到专业知识学习的过渡中起到承前启后的作用。本书主编结合自己在国外丰富的英语学习经验以及国内学生在生物专业英语学习方面的特点、难点，本着生物学各分支学科专业英语知识"比例均匀、难度适宜、注重实用、紧跟热点"的原则，选取了生物学各个分支学科的部分基础知识和热点研究内容进行编写并于2011年8月出版《生物专业英语》第一版。该教材因内容涵盖生物学各个分支学科的重要知识点，自面世以来被国内二十多所高等院校作为生物专业英语主讲教材使用，且得到一致好评。该教材已成为化学工业出版社畅销教材。

　　应出版社及各用书高校的邀请，主编在《生物专业英语》第一版基础上，根据用书高校学生反馈建议并针对生物学领域最新发展趋势，对部分内容进行了小幅修订，同时增加电子课件。《生物专业英语》第二版的编写在整体思路上仍然遵循第一版的原则和整体构架，教材选题仍然涵盖生物学（第1章）、微生物学（第2章）、细胞生物学（第3章）、植物学（第4章）、动物学（第5章）、分子遗传学（第6章）、生物化学（第7章）、生态学（第8章）、生物技术（第9章）与基因组学（第10章）等生物学以及其各个二级分支学科的专业英语知识，适合生物科学、生物工程、生物技术等各个专业学生用来进行专业英语学习。本教材内容较以往生物学专业英语教材更加丰富，而且难易适度，对于学生在科研领域更进一步深造具有很好的指引和铺垫作用。

　　主编结合自己在国内外科研、教学的丰富经验，精选生物学各个学科的国际研究热点作为学生的阅读材料，对于学生掌握科技前沿动态以及实用性专业英语知识具有重要意义。主编负责所有章节的选材、内容组织、翻译及练习题的编写工作，副主编及其他编写人员在选材以及个别章节内容的翻译、校对等方面做了大量的工作，本教材的主审胡宝忠教授以专业的眼光、开阔的思路高屋建瓴地提出了很多宝贵的建议，为本教材的最终成形提供了莫大的帮助。

回首为《生物专业英语》第一版和第二版倾注的心血和汗水，品味每个章节的内容，编者认为本教材具有以下特点：

1）专业英语内容选材丰富、知识点较新，各章节内容比例适当、难易程度适中。本教材的选材以各分支学科的英文原版教材为主，而且主要选取的是各分支学科的基础专业知识。本教材不仅涵盖了生物学及其各个分支学科的主要专业知识点，而且专业知识点完全以纯正英语方式表达，这使得学生更易于掌握已知专业知识点的英语表达内容和方式。

2）在讲解专业英语词汇时注重讲解它们的准确含义及其具体的使用语境。本教材通过在每篇文章中标注词汇的具体位置，使得学生在学习时能更快找到它们在文中所处的语境，更好地理解其含义及其使用方法。此外，在习题中设置"Matching"练习，使得学生通过英文讲解更好地理解并掌握较为重要或较常用的专业英语词汇的具体含义和使用方法。

3）补充阅读材料的选取多以各学科最新的研究进展或报道为主。在各章节主讲内容基础上，多以各分支学科的最新研究动态（或热点）为主，增加补充材料的趣味性和新颖性，这样不仅能通过阅读材料的选取有针对性地进一步强化各个章节的专业英语知识点，使得学生在学习专业英语知识后有目的地实践，而且可以增强学生阅读的兴趣，丰富自身的专业知识。

4）每章起始设置本章中文导读，每节设置重点疑难句子的中文翻译。考虑到学生在接触专业英语知识时不仅面临大量陌生的英语词汇，而且还不得不接受一些新的专业知识，因此，本教材在每章开篇通过"本章中文导读"对其各节知识点以及该章节的学习要点进行简要介绍，而且在每章节后通过"Notes to the Difficult Sentences"对重要的疑难专业英语知识（文中该疑难句子用下划线标出）进行翻译，这样可以使得学生通过中文导读和疑难句子的中文翻译对各章节的专业英语知识内容有一个更好的了解和掌握。

5）课后习题的多样化。考虑到学生在学习专业英语时刚刚结束大学英语的学习，由于他们在英语学习方面的惯性，往往需要一些习题对其学习效果进行强化和跟踪。本教材为各章节编排了丰富的习题。有为提高学生对专业英语词汇准确理解能力而设置的"Matching"练习，也有为提高学生专业英语的阅读能力而设置的"True or False"练习和"Reading Comprehension"练习，还有为提高学生专业英语翻译水平而设置的"Translation from English to Chinese"练习，以及为提高学生专业英语写作能力而设置的"Translation from Chinese to English"练习。这样，本教材以学生的学习特点和惯性作为编写习题的宗旨，强化学生的专业英语学习，更易于被学生接受，以达到提高学生专业英语水平的最终目的。

本教材的出版得到国家自然科学基金联合基金重点项目（编号 U23A20143）、面上项目（编号 32070031）和青年基金项目（编号 31000055 和 32000065）的资助。

编者
2024 年 5 月

CONTENTS

Chapter 1　Introduction to Biology ··· 1
　1.1　What is Biology? ··· 1
　1.2　The Origin of Life ··· 4
　1.3　The Significance of Biology in Your Life ··································· 8
　1.4　The History of Biology—Additional Reading ································ 11

Chapter 2　Microbiology ··· 15
　2.1　The Scope and Relevance of Microbiology ································· 15
　2.2　The Future of Microbiology ·· 21
　2.3　Prokaryotes, Eukaryotic Microbes and Viruses ····························· 26
　2.4　Extreme Microbes—Additional Reading ···································· 31

Chapter 3　Cellular Biology ·· 34
　3.1　The Discovery of Cells ··· 34
　3.2　Basic Properties of Cells ··· 37
　3.3　The Concepts in Mammalian Cell Culture—Additional Reading ············ 42

Chapter 4　Botany—Plant Biology ··· 49
　4.1　The Scope and Importance of Botany ······································ 49
　4.2　Flowers, Fruits and Seeds of Plants ·· 52
　4.3　Photosynthesis ·· 56
　4.4　Some Achievements of Transgenic Plants—Additional Reading ············ 59

Chapter 5　Zoology—Animal Biology ·· 65
　5.1　What is Zoology? ·· 65
　5.2　Tissues, Organs and Organ Systems of Animals ···························· 67
　5.3　Transgenic Animals—Additional Reading ·································· 72

Chapter 6　Molecular Genetics ··· 76
　6.1　Gene and Chromosomes ··· 76
　6.2　Functional Structure of a Gene ·· 79
　6.3　Gene Expression ·· 83
　6.4　Essentials for Genetic Engineering—Additional Reading ·················· 86

Chapter 7　Biochemistry ··· 91
　7.1　Enzymes ··· 91
　7.2　Metabolism ··· 93
　7.3　Energy Transformation ·· 96

    7.4   Protein crystallization—Additional Reading ························· 100

Chapter 8   Ecology ························· 104

    8.1   What is Ecology? ························· 104

    8.2   Ecosystems ························· 107

    8.3   Biodiversity—Additional Reading ························· 110

Chapter 9   Biotechnology ························· 114

    9.1   Biotechnology Overview ························· 114

    9.2   Recombinant DNA Technology ························· 118

    9.3   Recombinant Protein Expression ························· 121

    9.4   Bioalcohols—Additional Reading ························· 124

Chapter 10   Genomics ························· 128

    10.1   The First Sequenced Genomes ························· 128

    10.2   The Functional Genomics ························· 131

    10.3   Proteomics ························· 134

    10.4   Bioinformatics—Additional Reading ························· 137

Appendix Ⅰ   Answers to Exercises ························· 140

Appendix Ⅱ   Index of Professional Words and Phrases ························· 157

References ························· 171

Chapter 1　Introduction to Biology

[本章中文导读]

本章作为引言，主要介绍什么是生物学（第 1.1 节）、生命的起源（第 1.2 节）、生物学对人们的生活的重要性（第 1.3 节）以及作为补充阅读材料的生物学的发展史（第 1.4 节）等科普性的生物学专业知识。旨在通过对以上内容的英文讲解，使学生能够掌握一些生物学的基本专业词汇的具体含义以及相关的实际用法。

1.1　What is Biology?

Biology[1] is the study of life. Alongside physics and chemistry, biology is one of the largest and most important branches of science. At the highest level, biology is broken down based on the type of organism[2] being studied: zoology[3], the study of animals; botany[4], of plants; and microbiology[5], of microorganisms[6]. Each field has contributed to mankind or the Earth's well-being[7] in numerous ways. Most prominently: botany, to agriculture[8]; zoology, to livestock[9] and protection of ecologies[10]; and microbiology, to the study of disease[11] and ecosystems[12] in general.

Besides classifications[13] based on the category of organism being studied, biology contains many other specialized sub-disciplines, which may focus on just one category of organism or address organisms from different categories. This includes biochemistry[14], the interface between biology and chemistry; molecular biology[15], which looks at life on the molecular level[16]; cellular biology[17], which studies different types of cells and how they work; physiology[18], which looks at organisms at the level of tissues[19] and organs[20]; ecology, which studies the interactions[21] between organisms themselves; ethology[22], which studies the behavior of animals, especially complex animals; and genetics[23], overlapping[24] with molecular biology, which studies the code[25] of life, DNA.

The foundations of modern biology include four components: cell theory[26]; that life is made of fundamental units called cells; evolution[27], that life is not deliberately designed by rather evolves incrementally through random mutations[28] and natural selection[29]; gene theory[30], that tiny molecular sequences[31] of DNA dictate the entire structure of an organism and are passed from parents[32] to offspring[33]; and homeostasis[34], that each organism's body includes a complex suite of processes designed to preserve its biochemistry from the entropic effects[35] of the external environment.

The basic picture in biology has stayed roughly the same since DNA was first imaged using X-ray crystallography[36] in the 1950s, although there are constant refinements to the details, and life is so complex that it could be centuries or even millennia before we begin to understand it in its entirety. But it should be made clear that we are moving towards complete understanding: life, while complex, consists of a finite amount of complexity that only appreciably increases on relatively long timescales of hundreds of thousands or millions of years. Evolution, while creative, operates slowly. In recent years, much excitement in biology has centered on the sequencing of genomes[37] and their comparison, called genomics[38], and the creation of life with custom-written DNA programming, called synthetic biology[39]. These fields are sure to continue grabbing the headlines in the near future.

Notes to the Difficult Sentences

Besides classifications based on the category of organism being studied, biology contains many other specialized sub-disciplines, which may focus on just one category of organism or address organisms from different categories.

除了根据所研究生物种类的不同进行分类外，生物学还包括许多其他专门的分支学科，这些分支学科知识可能只关注一类生物，或者研究不同种类的生物。

The basic picture in biology has stayed roughly the same since DNA was first imaged using X-ray crystallography in the 1950s, although there are constant refinements to the details, and life is so complex that it could be centuries or even millennia before we begin to understand it in its entirety.

自从20世纪50年代利用X射线衍射得到DNA的结构以来，生物学的基本图景基本没变。尽管人们对生物学进行了不断探索，但生命现象如此复杂，以至于彻底研究清楚可能需要成百上千年的时间。

Professional Words and Phrases

[1] **biology** [baiˈɔlədʒi] n. 生物学
[2] **organism** [ˈɔːgənizəm] n. 生物体
[3] **zoology** [zəuˈɔlədʒi] n. 动物学
[4] **botany** [ˈbɔtəni] n. 植物学
[5] **microbiology** [maikrəubaiˈɔlədʒi] n. 微生物学
[6] **microorganism** [maikrəuˈɔːgənizəm] n. 微生物
[7] **well-being** [ˈwelˈbiːiŋ] n. 健康，福利
[8] **agriculture** [ˈægrikʌltʃə] n. 农业
[9] **livestock** [ˈlaivstɔk] n. 家畜，牲畜
[10] **ecology** [iːˈkɔlədʒi] n. 生态学
[11] **disease** [diˈziːz] n. 疾病
[12] **ecosystem** [ˈiːkəusistəm] n. 生态系统
[13] **classification** [klæsifiˈkeiʃən] n. 分类，类别
[14] **biochemistry** [ˈbaiəuˈkemistri] n. 生物化学
[15] **molecular biology** 分子生物学
[16] **molecular level** 分子水平
[17] **cellular biology** 细胞生物学

[18] **physiology** [fizi'ɔlədʒi] n. 生理学
[19] **tissue** ['tisju:] n. （生物的）组织
[20] **organ** ['ɔ:gən] n. （生物的）器官
[21] **interaction** [intə'rækʃən] n. 相互关系，相互作用
[22] **ethology** [i:'θɔlədʒi] n. 动物行为学
[23] **genetics** [dʒi'netiks] n. 遗传学
[24] **overlap** ['əuvə'læp] n. 重叠，重复
[25] **code** [kəud] n. 密码，法则
[26] **cell theory** n. 细胞理论
[27] **evolution** [i:və'lu:ʃən] n. 进化
[28] **random mutation** 随机突变
[29] **natural selection** 自然选择
[30] **gene theory** 基因理论
[31] **molecular sequence** 分子序列
[32] **parent** ['pɛərənt] n. 亲本，母本
[33] **offspring** ['ɔ:fspriŋ] n. 后代，子孙
[34] **homeostasis** [həumiə'steisis] n. 自动平衡，体内平衡
[35] **entropic effect** 熵效应
[36] **X-ray crystallography** X 射线晶体衍射学
[37] **genome** ['dʒi:nəum] n. 基因组
[38] **genomics** [dʒə'nəumiks] n. 基因组学
[39] **synthetic biology** 合成生物学

Exercises

1. Matching

1) biology   a) the study of animals
2) zoology   b) the interface between biology and chemistry
3) botany   c) the study of the behavior of animals
4) microbiology   d) the study of life
5) biochemistry   e) looks at life on the molecular level
6) molecular biology f) studies the code of life
7) ecology   g) the study of plants
8) ethology   h) the study of microorganisms
9) genetics   i) studies the interactions between organisms

2. True or False

1) Genetics is not related to molecular biology.
2) Each organism's body includes a complex suite of processes designed to preserve its biochemistry from the entropic effects of the external environment.
3) The evolution of life is deliberately designed, affected occasionally by random mutations and natural selection.
4) Physiology is used for studying the life on the molecular level.

5) Cellular biology is the study of the different types of cells and how they work.

3. Reading Comprehension

1) Which is not included in biology based on categories of a organism?

　A. zoology

　B. botany

　C. microbiology

　D. cell theory

2) What can be passed from parents to offspring?

　A. DNA

　B. protein

　C. RNA

　D. All of the above

3) Which technique was used for imaging of DNA structure?

　A. X-ray crystallography

　B. random mutation

　C. DNA sequencing

　D. DNA programming

4. Translation from English to Chinese

　　In recent years, much excitement in biology has centered on the sequencing of genomes and their comparison, called genomics, and the creation of life with custom-written DNA programming, called synthetic biology.

5. Translation from Chinese to English

　　生命是复杂的，但其复杂性相对而言是有限的，而且在成千上万年或是上百万年的漫长时间中仅仅是以非常微小的速度增加的。进化尽管是具有创造性的，然而进行起来又是非常缓慢的。

1.2　The Origin of Life

　　The origin of life[1] are thought to have occurred sometime between 4.4 billion years ago, when the oceans and continents were just starting to form, and 2.7 billion years ago, when it is widely accepted that microorganisms existed in vast numbers due to their influence over isotope[2] ratios in the relevant strata[3]. Where exactly in this 1.7 billion year range the true origin of life can be found is less certain. A controversial paper published in 2002 by the UCLA paleontologist William Schopf argued that wavy geological[4] formations called stromalites[5] in fact contain 3.5 billion year-old fossilized[6] algae[7] microbes[8]. Some paleontologists[9] disagree with Schopf's conclusions and estimate the first life at around 3.0 billion years of age instead of 3.5 billion.

　　Evidence from the Isua supracrustal[10] belt in Western Greenland suggests an even earlier date for the origin of life —3.85 billion years ago. S. Mojzis makes this estimate based on isotope concentrations. Because life preferentially uptakes the isotope Carbon-12[11], areas where life has existed contain a higher-than-normal ratio of Carbon-12 to its heavier isotope, Carbon-13. This is widely known, but the interpretation of sediments[12] is less straightforward, and paleontologists do not always agree on their colleague's conclusions.

We do not know the exact geological conditions[13] of this planet 3 billion years ago, but we do have a rough idea, and can recreate these conditions in a laboratory[14]. Stanley Miller and Harold Urey recreated these conditions in their famous 1953 investigation, the Miller-Urey experiment. Using a highly reduced (non-oxygenated[15]) mixture of gases such as methane[16], ammonia[17], and hydrogen[18], these scientists synthesized basic organic[19] monomers[20], such as amino acids[21], in a completely inorganic[22] environment. Now, free-floating amino acids are a far cry from self-replicating[23], metabolism[24]-imbued microorganisms, but they at least give a suggestion as to how things might have gotten started.

In the large warm oceans of early Earth, quintillions[25] of these molecules[26] would randomly collide and combine, eventually making a rudimentary[27] proto-genome[28] of some sort. However, this hypothesis[29] is confused by the fact that the environment created in the Miller-Urey experiment had high concentrations of chemicals that would have prevented the formation of complex polymers[30] from the monomer building blocks.

In the 1950s and 1960s, another researcher, Sidney Fox, made an early-Earth-like environment in a lab and studied the dynamics[31]. He observed the spontaneous formation of peptides[32] from amino acid precursors, and saw these chemicals sometimes arranged themselves into microspheres, or closed spherical membranes[33], which he suggested were protocells[34]. If certain microspheres[35] formed which were capable of encouraging the growth of additional microspheres around them, it would amount to a primitive[36] form of self-replication[37], and eventually Darwinian evolution[38] would take over, creating effective self-replicators like today's cyanobacteria[39].

Notes to the Difficult Sentences

Because life preferentially uptakes the isotope Carbon-12, areas where life has existed contain a higher-than-normal ratio of Carbon-12 to its heavier isotope, Carbon-13. This is widely known, but the interpretation of sediments is less straightforward, and paleontologists do not always agree on their colleague's conclusions.

因为生命优先吸收同位素碳-12，所以在生命存在的地方碳-12 与更重的同位素碳-13 的比值比正常值更高。这是众所周知的，但是用沉淀物来解释并不够直接，所以古生物学家们并不完全认同他们同行的结论。

Using a highly reduced (non-oxygenated) mixture of gases such as methane, ammonia, and hydrogen, these scientists synthesized basic organic monomers, such as amino acids, in a completely inorganic environment. Now, free-floating amino acids are a far cry from self-replicating, metabolism-imbued microorganisms, but they at least give a suggestion as to how things might have gotten started.

这些科学家使用高度还原的（非氧化的）甲烷、氨和氢等的气体混合物，在一个绝对无机环境中合成了氨基酸等基本的有机单体。尽管这些游离氨基酸远不同于现在的能自我复制并能进行各种新陈代谢的微生物，但是它们至少暗示了生命是怎样发生的。

Professional Words and Phrases

[1] **the origin of life**　生命的起源
[2] **isotope**　[ˈaisətəup]　n. 同位素
[3] **strata**　[ˈstreitə]　n. 地层
[4] **geological**　[dʒiəˈlɔdʒikəl]　adj. 地质学的
[5] **stromalite**　[strəuˈmætlait]　n. 叠层石
[6] **fossilize**　[ˈfɔsilaiz]　vt. & vi. 使成化石，使陈腐
[7] **algae**　[ˈældʒi:]　n. 水藻，海藻，alga 的复数形式
[8] **microbe**　[ˈmaikrəub]　n. 微生物
[9] **paleontologist**　[ˌpæliɔnˈtɔlədʒist]　n. 古生物学家
[10] **supracrustal**　[sju:prəˈkrʌstəl]　adj. (地层、岩组等)覆盖基底岩石的，上地壳的
[11] **carbon**　[ˈkɑ:bən]　n. 碳
[12] **sediment**　[ˈsedimənt]　n. 沉淀
[13] **geological condition**　地质条件
[14] **laboratory**　[ˈlæbrətɔ:ri]　n. 实验室
[15] **oxygenate**　[ˈɔksidʒineit]　v. 以氧处理，氧化；n. 氧化剂
[16] **methane**　[ˈmeθein]　n. 甲烷，沼气
[17] **ammonia**　[əˈməunjə]　n. 氨
[18] **hydrogen**　[ˈhaidrədʒən]　n. 氢
[19] **organic**　[ɔ:ˈgænik]　adj. 器官的，有机的
[20] **monomer**　[ˈmɔnəmə]　n. 单体
[21] **amino acid**　氨基酸
[22] **inorganic**　[ˈinɔ:ˈgænik]　adj. 无机的，无生物的
[23] **self-replicate**　[ˈselfˈreplikeit]　v. 自我复制
[24] **metabolism**　[meˈtæbəlizəm]　n. 新陈代谢
[25] **quintillion**　[kwinˈtiljən]　n. 百万的三次方
[26] **molecule**　[ˈmɔlikju:l]　n. 分子
[27] **rudimentary**　[ˌru:dəˈmentəri:]　adj. 基本的，初步的，未充分发展的
[28] **proto-genome**　n. 原始的基因组
[29] **hypothesis**　[haiˈpɔθəsis]　n. 假说，假设，猜测
[30] **polymer**　[ˈpɔlimə]　n. 多聚体
[31] **dynamics**　[daiˈnæmiks]　n. 力学，动力学，动态
[32] **peptide**　[ˈpeptaid]　n. 肽
[33] **membrane**　[ˈmembrein]　n. 薄膜，膜状物
[34] **protocell**　n. 细胞的原始状态
[35] **microsphere**　[ˈmaikrəsfiə]　n. 微球体，微滴
[36] **primitive**　[ˈprimitiv]　adj. 原始的，简陋的
[37] **self-replication**　n. 自我复制
[38] **Darwinian evolution**　达尔文进化
[39] **cyanobacteria**　[ˈsaiænəuˈbæktiə]　n. 蓝细菌，cyanobacterium 的复数形式

Exercises

1. Matching

1) stromalites   a) a scientist who studies paleontology, learning about the forms of life that existed in former geologic periods, chiefly by studying fossils
2) proto-genome   b) a proposed explanation for a phenomenon
3) paleontologist   c) the primitive form of a genome
4) protocell   d) wavy geological formations
5) hypothesis   e) the primitive form of a cell

2. True or False

1) The microorganisms existed in vast numbers due to their influence over isotope 1.7 billion years ago.
2) The true origin of life exactly started 3.5 billion years ago, just because wavy geological formations called stromalites in fact contain 3.5 billion year-old fossilized algae microbes.
3) The organic monomers could not be synthesized in a completely inorganic environment.
4) Miller and Urey made an early-Earth-like environment in a lab and studied the dynamics.
5) Sidney Fox suggested that the protocells were the microspheres or closed spherical membranes which chemicals such as peptides sometimes arranged themselves into.

3. Reading Comprehension

1) When was the origin of life thought to have occurred?
    A. 4.4 billion years ago
    B. 1.7 billion years ago
    C. between A and B
    D. 3.5 billion years ago

2) Which evidence is not used for the characterization of the origin of life?
    A. the fossilized algae microbes
    B. the ratio of Carbon-12 to Carbon-13
    C. the famous 1953 investigation
    D. cyanobacteria

3) What matter is involved in the highly reduced gases in the Miller-Urey experiment?
    A. methane
    B. ammonia
    C. hydrogen
    D. All

4. Translation from English to Chinese

     A controversial paper published in 2002 by the UCLA paleontologist William Schopf argued that wavy geological formations called stromalites in fact contain 3.5 billion year-old fossilized algae microbes. Some paleontologists disagree with Schopf's conclusions and estimate the first life at around 3.0 billion years of age instead of 3.5 billion.

5. Translation from Chinese to English

     如果某些微球体形成并能激发它们周围的其他微球体生长，那么它们就相当于一种原始的自我复制形式，并最终由达尔文进化接管这种生命形式，创造出像今天的蓝细菌这样的有效自我复制体。

1.3 The Significance of Biology in Your Life

Biology is the science that deals with the study of varieties of living organisms including ourselves. The significance of biology in your daily life lies in the fact that biology attempts to find out the unifying principle that exists among diverse organisms having morphological[1] and functional[2] inequalities[3]. The significance of biology in your daily life can be considered from the two natural divisions[4] of the science itself, plant life and animal life.

Agriculture plays great role in narrating the significance of biology in your daily life. Agriculture is largely the result of man's taking the advantage of the interrelations[5] of soil, climate and natural habitat[6] to select those particular combinations that meet his basic requirements. Thus to provide necessary food, man depends entirely on green plants[7] that can alone capture the solar energy[8]. High yielding varieties of crop plants like rice, wheat, jute[9], sugar cane[10], pulses[11] etc are now bred experimentally. Disease-resistant[12] grains[13] and vernalized[14] seeds are made. Biological control[15] strategies are undertaken as pest control[16] emphasizing the significance of biology in your daily life. Modern man does not depend on fishing and hunting like our ancestors and instead rears fishes as well as cattle and various other domestic[17] animals to get food and other necessities of life. This has resulted in the development of fishery[18] and animal husbandry[19]. The importance of biology in your daily lives lies in the production of clothes and timber for making furniture, in supplied raw materials for paper, dyes, etc. Fossils are important in locating underground oil and natural gas reserves. Even coal and mineral oil formed from decomposed[20] plant bodies are key to industrial prosperity.

Medical[21] advancement[22] also shows the significance of biology in your daily lives. The study of dreaded[23] diseases, their causative agents[24], cure as well as the actions of drugs are a way of biological enlightenment that strives minimizing human suffering. The significance of biology in your daily life also lies in finding and curing hereditary[25] abnormalities[26] like haemophilia[27], Down's syndrome[28], etc. Biology aims in making effort to better human race through eugenics[29]. Biology study has a vital role in controlling environmental pollution[30] and attracted sense of art and beauty.

Notes to the Difficult Sentences

Medical advancement also shows the significance of biology in your daily lives. The study of dreaded diseases, their causative agents, cure as well as the actions of drugs are a way of biological enlightenment that strives minimizing human suffering. The significance of biology in your daily life also lies in finding and curing hereditary abnormalities like haemophilia, Down's syndrome, etc. Biology aims in making effort to better human race through eugenics.

医学的进步也显示了生物学在日常生活中的重要意义。研究可怕的疾病、致病原因、治疗疾病的方法以及药物活性是人类领悟生物学努力使人类痛苦最小化的一种方式。生物学对日常生活的重要意义也体现在人们对血友病、唐氏综合征等遗传异常疾病的发现和治疗方面。生物学的目标就是借助优生学知识努力改善人类种族的健康状况。

Professional Words and Phrases

[1] **morphological** [mɔ:fə'lɔdʒikəl] adj. 形态学的
[2] **functional** ['fʌŋkʃənəl] adj. 功能的，实用的
[3] **inequality** [ini'kwɔləti] n. 不平等，不平均，不平坦
[4] **division** [di'viʒən] n. 划分，部门
[5] **interrelation** [intəri'leiʃən] n. 相互关系
[6] **habitat** ['hæbitæt] n. 栖息地，产地
[7] **green plants** 绿色植物
[8] **solar energy** 太阳能
[9] **jute** [dʒu:t] n. 黄麻
[10] **sugar cane** 甘蔗
[11] **pulse** [pʌls] n. 豆类等结荚植物可食性种子
[12] **disease-resistant** adj. 抗病的
[13] **grain** [grein] n. 谷物，谷类
[14] **vernalize** ['və:nəlaiz] v. 施以春化处理，以人工方法促进发育
[15] **biological control** 生物学防治
[16] **pest control** 昆虫防治
[17] **domestic** [də'mestik] adj. 驯养的
[18] **fishery** ['fiʃəri] n. 渔场，渔业
[19] **animal husbandry** 畜牧业
[20] **decompose** [di:kəm'pəuz] vi. 分解，腐烂; vt. 腐烂
[21] **medical** ['medikəl] adj. 医疗的，医学的，医药的，内科的
[22] **medical advancement** 医学进展
[23] **dread** [dred] adj. 可怕的，恐怖的; v. 恐惧，害怕
[24] **causative agent** 病原体，病原物
[25] **hereditary** [hi'reditəri] adj. 世袭的，遗传的
[26] **abnormality** [æbnɔ:'mæliti] n. 异常，畸形
[27] **haemophilia** [hi:mə'filiə] n. 血友病，出血不止
[28] **Down's syndrome** 唐氏综合征
[29] **eugenics** [ju:'dʒeniks] n. 优生学
[30] **environmental pollution** 环境污染

Exercises

1. Matching

1) agriculture   a) genetically transmitted or transmittable from parent to offspring
2) morphological   b) a group of hereditary genetic disorders that impair the body's ability to control blood clotting or coagulation
3) hereditary   c) the cultivation of animals, plants, fungi and other life forms for food,

4) haemophilia	d) the applied science or the biosocial movement which advocates the use of practices aimed at improving the genetic composition of a population
5) decompose	e) relating to or concerned with the morphology of plants, microorganisms and animals
6) eugenics	f) to separate into components or basic elements or cause to rot

(continued from previous: fiber, and other products used to sustain life)

2. True or False

1) Biology attempts to find out the unifying principle that exists among diverse organisms having morphological and functional inequalities.

2) Our ancestors did not depend on fishing and hunting.

3) Biology aims in making effort to better human race through agriculture.

4) Fossils are not important in locating underground oil and natural gas reserves.

5) Biological control is largely the result of man's taking the advantage of the interrelations of soil, climate and natural habitat.

3. Reading Comprehension

1) Which does not explain the significance of biology in our life?
   A. To find out the unifying principle that exists among diverse organisms
   B. Biological control strategies are undertaken as pest control
   C. Medical advancement
   D. None of the above

2) Which is not true of the following descriptions about agriculture?
   A. Agriculture strives minimizing human suffering.
   B. Agriculture can provide mankind with necessary food.
   C. Disease-resistant grains and vernalized seeds are made through the development of agriculture.
   D. Agriculture helps modern man better their life conditions, as compared with their ancestors.

3) Which is not related to the medical advancements?
   A. The study of dreaded diseases
   B. The study of the causative agents
   C. The study of the actions of drugs
   D. Environmental pollution

4. Translation from English to Chinese

The significance of biology in your daily life lies in the fact that biology attempts to find out the unifying principle that exists among diverse organisms having morphological and functional inequalities. The significance of biology in your daily life can be considered from the two natural divisions of the science itself, plant life and animal life.

5. Translation from Chinese to English

农业在很大程度上是人类利用土壤、气候和自然环境的相互关系，选择其特定组合来满足人类自身基本需求的产物。因此为了提供必要的食物，人类必须完全依赖于能独自获取太阳能的绿色植物。像大米、小麦、黄麻、甘蔗以及豆类等各种高产农作物现在已经可以通过实验培育出来。抗病的谷类和春化过的种子已经被制造出来。

1.4 The History of Biology—Additional Reading

The history of biology traces the study of the living world from ancient to modern times. Although the concept of biology as a single coherent field arose in the 19th century, the biological sciences emerged from traditions of medicine and natural history reaching back to ayurveda, ancient Egyptian medicine[1] and the works of Aristotle and Galen in the ancient Greco-Roman world.

This ancient work was further developed in the Middle Ages by Muslim physicians and scholars such as Avicenna. During the European Renaissance and early modern period, biological thought was revolutionized in Europe by a renewed interest in empiricism and the discovery of many novel[2] organisms. Prominent in this movement were Vesalius and Harvey, who used experimentation[3] and careful observation in physiology, and naturalists such as Linnaeus and Buffon who began to classify[4] the diversity[5] of life and the fossil record, as well as the development and behavior of organisms. Microscopy[6] revealed the previously unknown world of microorganisms, laying the groundwork for cell theory. The growing importance of natural theology[7], partly a response to the rise of mechanical philosophy[8], encouraged the growth of natural history (although it entrenched the argument from design).

Over the 18th and 19th centuries, biological sciences[9] such as botany and zoology became increasingly professional scientific disciplines. Lavoisier and other physical scientists began to connect the animate and inanimate worlds through physics and chemistry. Explorer -naturalists[10] such as Alexander von Humboldt investigated the interaction between organisms and their environment, and the ways this relationship depends on geography laying the foundations for biogeography[11], ecology and ethology. Naturalists began to reject essentialism and consider the importance of extinction and the mutability[12] of species[13]. Cell theory provided a new perspective on the fundamental basis of life. These developments, as well as the results from embryology[14] and paleontology, were synthesized in Charles Darwin's theory of evolution by natural selection. The end of the 19th century saw the fall of spontaneous generation[15] and the rise of the germ[16] theory of disease, though the mechanism[17] of inheritance[18] remained a mystery.

In the early 20th century, the rediscovery of Mendel's work led to the rapid development of genetics by Thomas Hunt Morgan and his students, and by the 1930s the combination of population genetics[19] and natural selection in the "neo-Darwinian synthesis". New disciplines developed rapidly, especially after Watson and Crick proposed the structure of DNA. Following the establishment of the Central Dogma[20] and the cracking of the genetic code[21], biology was largely split between organismal biology[22]—the fields that deal with whole organisms and groups of organisms—and the fields related to cellular and molecular biology. By the late 20th century, new fields like genomics and proteomics[23] were reversing this trend, with organismal biologists[24] using molecular techniques[25], and molecular and cell biologists investigating the interplay between genes and the environment, as well as the genetics of natural populations of organisms.

At the beginning of the 21st century, biological sciences converged with previously differentiated new and classic disciplines like Physics into research fields like biophysics[26]. Advances were made in analytical chemistry[27] and physics instrumentation[28] including improved sensors[29], optics[30], tracers, instrumentation, signal processing, networks, robots, satellites, and compute power for data collection, storage, analysis, modeling, visualization, and simulations. These technology advances allowed theoretical and experimental research including internet publication of molecular biochemistry[31], biological systems[32], and ecosystems science. This enabled worldwide access to better measurements, theoretical models, complex simulations, theory predictive model experimentation, analysis, worldwide internet observational data reporting, open peer-review, collaboration, and internet publication. New fields of biological sciences research emerged including bioinformatics[33], theoretical biology[34], computational genomics[35], astrobiology[36] and synthetic biology.

Notes to the Difficult Sentences

Explorer-naturalists such as Alexander von Humboldt investigated the interaction between organisms and their environment, and the ways this relationship depends on geography laying the foundations for biogeography, ecology and ethology. Naturalists began to reject essentialism and consider the importance of extinction and the mutability of species. Cell theory provided a new perspective on the fundamental basis of life. These developments, as well as the results from embryology and paleontology, were synthesized in Charles Darwin's theory of evolution by natural selection. The end of the 19th century saw the fall of spontaneous generation and the rise of the germ theory of disease, though the mechanism of inheritance remained a mystery.

像 Alexander von Humboldt 等探险家兼博物学家研究了生物体与其环境间的相互作用及其依赖于地理学的方式，这就为生物地理学、生态学和动物行为学奠定了基础。博物学家开始抵制实在论并思考物种灭绝和易变性的重要性。细胞理论提供了一个观察生命理论基础的新视角。这些发展以及胚胎学和古生物学的研究成果是在以自然选择为基础的达尔文进化理论中综合起来的。十九世纪末见证了自然发生说的衰落和病原体理论的兴起，尽管遗传机制仍然是个谜。

Professional Words and Phrases

[1] **medicine** ['medisin] n. 医学
[2] **novel** ['nɔvəl] adj. 新奇的
[3] **experimentation** [iksperimen'teiʃn] n. 实验方法
[4] **classify** ['klæsifai] vt. 分类，归类
[5] **diversity** [dai'və:siti] n. 多样性，差异
[6] **microscopy** ['maikrəskəupi] n. 显微镜使用，用显微镜检查
[7] **theology** [θi'ɔlədʒi] n. 神学
[8] **mechanical philosophy** 机械论哲学
[9] **biological sciences** 生物科学

[10] **naturalist** [ˈnætʃərəlist] n. 博物学家，自然主义者
[11] **biogeography** [baiəudʒiˈɔgrəfi] n. 生物地理学
[12] **mutability** [mju:təˈbiləti] n. 易变性
[13] **species** [ˈspi:ʃi:z] n. 种类，（单复同）物种
[14] **embryology** [embriˈɔlədʒi] n. 胚胎学，发生学
[15] **spontaneous generation** 自然发生
[16] **germ** [dʒə:m] n. 病原微生物
[17] **mechanism** [ˈmekənizəm] n. 机制，原理
[18] **inheritance** [inˈheritəns] n. 遗传，继承
[19] **population genetics** 群体遗传学
[20] **Central Dogma** 中心法则
[21] **genetic code** 遗传密码
[22] **organismal biology** 有机体生物学
[23] **proteomics** [prəutiˈəumiks] n. 蛋白质组学
[24] **biologist** [baiˈɔlədʒist] n. 生物学家
[25] **molecular technique** 分子技术
[26] **biophysics** [baiəuˈfiziks] n. 生物物理学
[27] **analytical chemistry** 分析化学
[28] **instrumentation** [instrəmenˈteiʃən] n. 仪器、仪表
[29] **sensor** [ˈsensə] n. 传感器
[30] **optics** [ˈɔptiks] n. 光学
[31] **molecular biochemistry** 分子生物化学
[32] **biological systems** 生物系统
[33] **bioinformatics** [baiɔinfəˈmætiks] n. 生物信息学
[34] **theoretical biology** 理论生物学
[35] **computational genomics** 计算基因组学
[36] **astrobiology** [æstrəˈbaiɔdʒi] n. 天体生物学

Exercises

1. Matching

1) novel   a) the set of rules by which information encoded in genetic material (DNA or mRNA sequences) is translated into proteins (amino acid sequences) by living cells

2) experimentation   b) deals with the detailed residue-by-residue transfer of sequential information

3) theology   c) one that advocates or practices naturalism

4) naturalist   d) the systematic study of religion and its influences and of the nature of supposed religious truths, or the learned profession acquired by completing specialized courses in religion, usually at a college or seminary

5) species   e) the step in the scientific method that helps people decide between two

or more competing explanations – or hypotheses

6) Central Dogma   f) one of the basic units of biological classification and a taxonomic rank.

7) genetic code   g) original or striking especially in conception or style

2. True or False

1) The concept of biology as a single coherent field arose in the 18th century.

2) Avicenna was a famous scholar in the ancient Greco-Roman world.

3) Biogeography, ecology and ethology resulted from the study of the interaction between organisms and their environment, and the ways this relationship depends on geography.

4) Thomas Hunt Morgan first proposed the structure of DNA.

5) Organismal biology is the field related to cellular and molecular biology.

3. Reading Comprehension

1) Which technique revealed the previously unknown world of microorganisms?

   A. careful observation in physiology

   B. experimentation

   C. both A and B

   D. microscopy

2) Which of the statements below is not correct?

   A. The spontaneous generation conflict came up at the end of the 19th century.

   B. The germ theory of disease arose at the end of the 19th century.

   C. The mechanisms of inheritance was known at the beginning of the 19th century.

   D. Over the 18th and 19th centuries, biological sciences became increasingly professional scientific disciplines.

3) What is not included in the new fields of biological sciences research of the 21st century?

   A. bioinformatics

   B. theoretical biology

   C. synthetic biology

   D. proteomics

4. Translation from English to Chinese

New disciplines developed rapidly, especially after Watson and Crick proposed the structure of DNA. Following the establishment of the Central Dogma and the cracking of the genetic code, biology was largely split between organismal biology—the fields that deal with whole organisms and groups of organisms—and the fields related to cellular and molecular biology. By the late 20th century, new fields like genomics and proteomics were reversing this trend, with organismal biologists using molecular techniques, and molecular and cell biologists investigating the interplay between genes and the environment, as well as the genetics of natural populations of organisms.

5. Translation from Chinese to English

所取得的进展表现在分析化学方面和包括改进的传感器、光学、示踪剂、仪器、信号处理、网络、机器人、卫星在内的物理仪器方面，以及用于数据集成、储存、分析、建模、可视化、模拟的计算能力方面。这些技术进步使得在网上发表分子生物化学、生物系统和生态系统学等方面的理论和实验研究成果成为可能。

Chapter 2   Microbiology

[本章中文导读]

  微生物被认为是我们这个星球上首先发现的生命体，几乎存在于目前已知的任何地方。微生物学是研究微生物及其相关知识的学科，是生物学最重要的分支学科之一。本章主要介绍了微生物学的范围及相关学科（第 2.1 节），微生物学的未来（第 2.2 节），原核微生物、真核微生物和病毒（第 2.3 节），以及作为补充阅读材料的极端微生物（第 2.4 节）等微生物学专业知识。作者选取以上内容主要是想通过对以上知识点的英文讲解，使学生掌握有关的微生物学中重要的专业英语词汇的具体含义及其实际用法等。

2.1   The Scope and Relevance of Microbiology

  As the scientist-writer Steven Jay Gould emphasized, we live in the Age of Bacteria[1]. They are the first living organisms on our planet, live virtually everywhere life is possible, are more numerous than any other kind of organism, and probably constitute the largest component of the earth's biomass[2]. The whole ecosystem depends on their activities, and they influence human society in countless ways. Thus modern microbiology[3] is a large discipline with many different specialties[4]; it has a great impact on fields such as medicine, agricultural and food science[5], ecology, genetics, biochemistry, and molecular biology.

  For example, microbiology has been a major contributor to the rise of molecular biology, the branch of biology dealing with the physical and chemical aspects of living matter and its function. Microbiologists[6] have been deeply involved in studies on the genetic code and the mechanisms of DNA, RNA, and protein synthesis. Microorganisms were used in many of the early studies on the regulation of gene expression[7] and the control of enzyme activity[8]. In the 1970s new discoveries in microbiology led to the development of recombinant DNA technology[9] and genetic engineering[10].

  One indication[11] of the importance of microbiology in the twentieth century is the Nobel Prize given for work in physiology or medicine. About 1/3 of these have been awarded to scientists working on microbiological problems. Microbiology has both basic and applied aspects. Many microbiologists are interested primarily in the biology of the microorganisms themselves. They may focus on a specific group of microorganisms and be called virologists[12] (viruses[13]), bacteriologists[14] (bacteria), phycologists[15] or algologists[16] (algae), mycologists[17] (fungi[18]), or protozoologists[19] (protozoa[20]). Others are interested in microbial morphology[21] or particular functional processes and work in fields such as microbial cytology[22], microbial physiology[23], microbial ecology[24], microbial genetics[25] and molecular biology, and microbial taxonomy[26]. Of course a person can be thought of in both ways (e.g., as a bacteriologist who works on taxonomic[27] problems). Many microbiologists have a more applied orientation and work on practical problems in fields such as medical microbiology[28], food and dairy microbiology[29], and public health microbiology[30] (basic research is also conducted in these fields). Because the various fields of microbiology are interrelated, an applied microbiologist[31] must be familiar with basic microbiology[32]. For example, a medical microbiologist[33] must have a good understanding of microbial taxonomy, genetics, immunology[34], and physiology to identify and properly respond to the pathogen[35] of concern.

What are some of the current occupations of professional microbiologists?

One of the most active and important is medical microbiology, which deals with the diseases of humans and animals. Medical microbiologists identify the agent causing an infectious disease[36] and plan measures to eliminate it. Frequently they are involved in tracking down new, unidentified[37] pathogens such as the agent that causes variant Creutzfeldt-Jacob disease[38], the hantavirus[39], and the virus responsible for AIDS[40]. These microbiologists also study the ways in which microorganisms cause disease.

Public health microbiology is closely related to medical microbiology. Public health microbiologists[41] try to control the spread of communicable diseases. They often monitor community food establishments and water supplies in an attempt to keep them safe and free from infectious disease agents.

Immunology is concerned with how the immune system[42] protects the body from pathogens and the response of infectious agents[43]. It is one of the fastest growing areas in science; for example, techniques for the production and use of monoclonal antibodies[44] have developed extremely rapidly. Immunology also deals with practical health problems such as the nature and treatment of allergies[45] and autoimmune diseases[46] like rheumatoid arthritis[47].

Many important areas of microbiology do not deal directly with human health and disease but certainly contribute to human welfare. Agricultural microbiology[48] is concerned with the impact of microorganisms on agriculture. Agricultural microbiologists[49] try to combat plant diseases that attack important food crops, work on methods to increase soil fertility and crop yields, and study the role of microorganisms living in the digestive tracts of ruminants[50] such as cattle. Currently there is great interest in using bacterial[51] and viral[52] insect[53] pathogens as substitutes for chemical pesticides[54].

The field of microbial ecology is concerned with the relationships between microorganisms and their living and nonliving habitats. Microbial ecologists[55] study the contributions of microorganisms to the carbon, nitrogen[56], and sulfur[57] cycles in soil and in freshwater. The study of pollution effects on microorganisms is also important because of the impact these organisms have on the environment. Microbial ecologists are employing microorganisms in bioremediation[58] to reduce pollution effects.

Scientists working in food and dairy microbiology try to prevent microbial spoilage of food and the transmission of food-borne[59] diseases such as botulism[60] and salmonellosis[61]. They also use microorganisms to make foods such as cheeses[62], yogurts[63], pickles[64], and beer. In the future microorganisms themselves may become a more important nutrient source[65] for livestock and humans.

In industrial microbiology[66] microorganisms are used to make products such as antibiotics[67], vaccines[68], steroids[69], alcohols[70] and other solvents[71], vitamins[72], amino acids, and enzymes[73]. Microorganisms can even leach valuable minerals from low-grade ores.

Research on the biology of microorganisms occupies the time of many microbiologists and also has practical applications. Those working in microbial physiology and biochemistry study the synthesis of antibiotics and toxins[74], microbial energy[75] production, the ways in which microorganisms survive harsh environmental conditions, microbial nitrogen fixation[76], the effects of chemical and physical agents on microbial growth[77] and survival[78], and many other topics.

<u>Microbial genetics[79] and molecular biology focus on the nature of genetic information[80] and how it regulates the development and function of cells and organisms. The use of microorganisms has been very helpful in understanding gene function. Microbial geneticists[81] play an important role in applied microbiology by producing new microbial strains[82] that are more efficient in synthesizing useful products.</u> Genetic techniques are used to test substances for their ability to cause cancer. More recently the field of genetic engineering has arisen from work in microbial genetics and molecular biology and will contribute substantially to microbiology, biology as a whole, and medicine. Engineered microorganisms are used to make hormones[83], antibiotics, vaccines, and other products. New genes can be inserted into plants and animals; for example, it may be possible to give corn and wheat nitrogen-fixation genes so they will not require nitrogen fertilizers.

Notes to the Difficult Sentences

Immunology is concerned with how the immune system protects the body from pathogens and the response of infectious agents. It is one of the fastest growing areas in science; for example, techniques for the production and use of monoclonal antibodies have developed extremely rapidly. Immunology also deals with practical health problems such as the nature and treatment of allergies and autoimmune diseases like rheumatoid arthritis.

免疫学关注的是免疫系统怎样保护机体免受病原体侵害以及如何响应传染源。它是科学中发展最快的领域；例如，单克隆抗体的制备及使用技术已经取得了极快速的发展。免疫学也负责研究诸如变态反应和类风湿性关节炎这样的自身免疫疾病的本质和治疗等实际健康问题。

Microbial genetics and molecular biology focus on the nature of genetic information and how it regulates the development and function of cells and organisms. The use of microorganisms has been very helpful in understanding gene function. Microbial geneticists play an important role in applied microbiology by producing new microbial strains that are more efficient in synthesizing useful products.

微生物遗传学和分子生物学主要研究的是遗传信息的本质以及它是如何调节细胞以及生物体的发育和功能的。微生物的利用对我们理解基因的功能是非常有帮助的。微生物遗传学家通过制造更为有效地合成有用产物的微生物新菌株，对微生物学的应用起到了重要作用。

Professional Words and Phrases

[1]　**bacteria**　[bæk'tiriə]　n. 细菌, bacterium 的复数形式
[2]　**biomass**　['baiəu'mæs]　n. 生物量

[3]　**modern microbiology**　现代微生物学
[4]　**specialty**　[ˈspeʃəlti]　n. 专业，专长
[5]　**food science**　食品科学
[6]　**microbiologist**　[maikrəubaiˈɔləgist]　n. 微生物学家
[7]　**gene expression**　基因表达
[8]　**enzyme activity**　酶活性
[9]　**recombinant DNA technology**　重组 DNA 技术
[10]　**genetic engineering**　遗传工程
[11]　**indication**　[indiˈkeiʃən]　n. 指示，表示，迹象
[12]　**virologist**　[vaiˈrɔləgist]　n. 病毒学家
[13]　**virus**　[ˈvairəs]　n. 病毒，病原体
[14]　**bacteriologist**　[bæktiriəˈɔlədʒist]　n. 细菌学家
[15]　**phycologist**　[faiˈkɔlədʒist]　n. 藻类学家
[16]　**algologist**　[ælˈgɔlədʒist]　n. 藻类学家
[17]　**mycologist**　[maiˈkɔlədʒist]　n. 真菌学家
[18]　**fungi**　[ˈfʌndʒai]　n. 真菌，fungus 的复数形式
[19]　**protozoologist**　[prəutəzəuˈɔlədʒist]　n. 原生动物学家
[20]　**protozoa**　[prəutəˈzəuə]　n. 原生动物，protozoan 的复数形式
[21]　**microbial morphology**　微生物形态学
[22]　**microbial cytology**　微生物细胞学
[23]　**microbial physiology**　微生物生理学
[24]　**microbial ecology**　微生物生态学
[25]　**microbial genetics**　微生物遗传学
[26]　**microbial taxonomy**　微生物分类学
[27]　**taxonomic**　[tæksəˈnɔmik]　adj. 分类学的
[28]　**medical microbiology**　医学微生物学
[29]　**dairy microbiology**　乳品微生物学
[30]　**public health microbiology**　公共健康微生物学
[31]　**applied microbiologist**　应用微生物学家
[32]　**basic microbiology**　基础微生物学
[33]　**medical microbiologist**　医学微生物学家
[34]　**immunology**　[imjuˈnɔlədʒi]　n. 免疫学
[35]　**pathogen**　[ˈpæθədʒən]　n. 病原体
[36]　**infectious disease**　传染病
[37]　**unidentified**　[ˈʌnaiˈdentifaid]　adj. 未确认的，无法识别的
[38]　**Creutzfeldt-Jacob disease**　克-雅病
[39]　**hantavirus**　[ˈhæntəˈvairəs]　n. 汉坦病毒
[40]　**AIDS**　[eidz]　abbr. 艾滋病，获得性免疫缺陷综合征(acquired immune deficiency syndrome)
[41]　**public health microbiologist**　公共健康微生物学家
[42]　**immune system**　免疫系统

[43] **infectious agents** 传染源，传染物
[44] **monoclonal antibody** 单克隆抗体
[45] **allergy** ['ælədʒi] n. 变态反应
[46] **autoimmune disease** 自身免疫病
[47] **rheumatoid arthritis** 类风湿性关节炎
[48] **agricultural microbiology** 农业微生物学
[49] **agricultural microbiologist** 农业微生物学家
[50] **ruminant** ['ru:minənt] n. 反刍动物
[51] **bacterial** [bæk'tiriəl] adj. 细菌的，细菌性的
[52] **viral** ['vairəl] adj. 病毒的
[53] **insect** ['insekt] n. 昆虫，虫子
[54] **pesticide** ['pestisaid] n. 杀虫剂，农药
[55] **microbial ecologist** 微生物生态学家
[56] **nitrogen** ['naitrədʒən] n. 氮
[57] **sulfur** ['sʌlfə] n. 硫
[58] **bioremediation** [baiɔˈriːmiːdiˈeiʃən] n. 生物修复
[59] **food-borne** ['fu:d bɔ:n] adj. 食物传染的
[60] **botulism** ['bɔtʃəlizəm] n. 肉毒杆菌中毒
[61] **salmonellosis** [sælmənəˈləusis] n. 沙门氏菌病
[62] **cheese** [tʃi:z] n. 奶酪
[63] **yogurt** ['jəugə:t] n. 酸奶
[64] **pickle** ['pikəl] n. 泡菜
[65] **nutrient source** 营养源
[66] **industrial microbiology** 工业微生物学
[67] **antibiotics** [æntibaiˈɔtiks] n. 抗生素
[68] **vaccine** ['væksi:n] n. 疫苗
[69] **steroid** ['stiərɔid] n. 类固醇
[70] **alcohol** ['ælkəhɔl] n. 醇类，乙醇
[71] **solvent** ['sɔlvənt] n. 溶剂
[72] **vitamin** ['vaitəmin] n. 维生素
[73] **enzyme** ['enzaim] n. 酶
[74] **toxin** ['tɔksin] n. 毒素
[75] **microbial energy** 微生物能源
[76] **nitrogen fixation** 固氮作用
[77] **microbial growth** 微生物生长
[78] **survival** [səˈvaivəl] n. 生存，存活
[79] **microbial genetics** 微生物遗传学
[80] **genetic information** 遗传信息
[81] **microbial geneticist** 微生物遗传学家
[82] **strain** [strein] n. 菌株
[83] **hormone** ['hɔ:məun] n. 荷尔蒙，激素

Exercises

1. Matching

1) biomass — a) a broad branch of biomedical science that covers the study of all aspects of the immune system in all organisms
2) virus — b) a microbe or microorganism such as a virus, bacterium, prion, or fungus that causes disease in its animal or plant host
3) bacteriologist — c) biological material from living, or recently living organisms
4) immunology — d) a small infectious agent that can replicate only inside the living cells of organisms
5) pesticide — e) microbiologists who focus mainly on bacteria
6) allergy — f) a biological preparation that improves immunity to a particular disease
7) bioremediation — g) proteins that catalyze chemical reactions
8) pathogen — h) any substance or mixture of substances intended for preventing, destroying, repelling or mitigating any pest
9) vaccine — i) the use of microorganism metabolism to remove pollutants
10) enzyme — j) a hypersensitivity disorder of the immune system

2. True or False

1) The field of microbial ecology is concerned with the relationships between microorganisms and their living and nonliving habitats.
2) In agricultural microbiology microorganisms are used to make products such as antibiotics, vaccines, steroids, alcohols and other solvents, vitamins, amino acids, and enzymes.
3) Medical microbiology is use to deal with the diseases of humans and animals.
4) Recently the field of genetic engineering has arisen from work in microbial genetics and molecular biology.
5) Biochemistry deals with practical health problems such as the nature and treatment of allergies and autoimmune diseases like rheumatoid arthritis.
6) Microbial ecologists are employing microorganisms in bioremediation to reduce pollution effects.
7) All microbiologists must only focus on a specific group of microorganisms such as viruses, bacteria, phycologists or algae, fungi, or protozoa.

3. Reading Comprehension

1) Which is not largely affected by modern microbiology?
   A. medicine
   B. ecology
   C. agricultural and food microbiology
   D. physics

2) Which of the following statements is not related to medical microbiologists?
   A. to identify the agent causing an infectious disease and plan measures to eliminate it
   B. to be concerned with the impact of microorganisms on agriculture
   C. to be involved in tracking down new, unidentified pathogens

D. study the ways in which microorganisms cause disease

3) Which is not an unidentified pathogen?

A. the agent that causes variant Creutzfeldt-Jacob disease

B. the hantavirus

C. the virus responsible for AIDS

D. algae

4) What is not studied by microbial ecologists?

A. the contributions of microorganisms to the carbon, nitrogen, and sulfur cycles

B. the effects of pollution on microorganisms

C. employing microorganisms in bioremediation to reduce pollution effects

D. the ways in which microorganisms cause disease

5) Which of the statements below is not true?

A. About 1/3 of the Nobel Prizes given for work in physiology or medicine have been awarded to scientists working on microbiological problems in the 20th century.

B. Agricultural microbiologists try to prevent microbial spoilage of food and the transmission of food-borne diseases such as botulism and salmonellosis.

C. Public health microbiology is closely related to medical microbiology.

D. An applied microbiologist must be familiar with basic microbiology.

4. Translation from English to Chinese

As the scientist-writer Steven Jay Gould emphasized, we live in the Age of Bacteria. They were the first living organisms on our planet, live virtually everywhere life is possible, are more numerous than any other kind of organism, and probably constitute the largest component of the earth's biomass. The whole ecosystem depends on their activities, and they influence human society in countless ways. Thus modern microbiology is a large discipline with many different specialties; it has a great impact on fields such as medicine, agricultural and food sciences, ecology, genetics, biochemistry, and molecular biology.

5. Translation from Chinese to English

许多微生物学家主要致力于微生物的生物学研究，以及微生物实际应用性研究。从事微生物生理学和生物化学研究的微生物学家研究的是抗生素和毒素的合成、微生物能源的产生、微生物在恶劣环境中的生存方式、微生物固氮作用、化学和物理因子对微生物生长和生存的作用以及许多其他课题。

2.2 The Future of Microbiology

As it is known, microbiology has had a profound influence on society. What of the future? Science writer Bernard Dixon is very optimistic about microbiology's future for two reasons. First, microbiology has a clearer mission than do many other scientific disciplines. Second, it is confident of its value because of its practical significance. Dixon notes that microbiology is required both to face the threat of new and reemerging human infectious diseases and to develop industrial technologies that are more efficient and environmentally friendly.

What are some of the most promising areas for future microbiological research and their potential practical impacts? What kinds of challenges do microbiologists face? The following brief list should give some idea of what the future may hold:

(1) New infectious diseases are continually arising and old diseases are once again becoming widespread and destructive. AIDS, hemorrhagic[1] fevers[2], and tuberculosis are excellent examples of new and reemerging infectious diseases. Microbiologists will have to respond to these threats, many of them presently unknown.

(2) <u>Microbiologists must find ways to stop the spread of established infectious diseases. Increases in antibiotic[3] resistance will be a continuing problem, particularly the spread of multiple drug resistance[4] that can render a pathogen impervious[5] to current medical treatment. Microbiologists have to create new drugs and find ways to slow or prevent the spread of drug resistance. New vaccines must be developed to protect against diseases such as AIDS. It will be necessary to use techniques in molecular biology and recombinant DNA technology to solve these problems.</u>

(3) Research is needed on the association between infectious agents and chronic diseases[6] such as autoimmune[7] and cardiovascular diseases[8]. It may be that some of these chronic afflictions[9] partly result from infections.

(4) We are only now beginning to understand how pathogens interact with host cells[10] and the ways in which diseases arise. There also is much to learn about how the host[11] resists pathogen invasions[12].

(5) Microorganisms are increasingly important in industry and environmental control[13], and we must learn how to use them in a variety of new ways. For example, microorganisms can (a) serve as sources of high-quality food and other practical products such as enzymes for industrial applications, (b) degrade pollutants[14] and toxic wastes[15], and (c) be used as vectors to treat diseases and enhance agricultural productivity[16]. There is also a continuing need to protect food and crops from microbial damage.

(6) Microbial diversity[17] is another area requiring considerable research. Indeed, it is estimated that less than 1% of the earth's microbial population has been cultured. We must develop new isolation techniques and an adequate classification of microorganisms, one which includes those microbes that cannot be cultivated in the laboratory. Much work needs to be done on microorganisms living in extreme environments[18]. The discovery of new microorganisms may well lead to further advances in industrial processes and enhanced environmental control.

(7) Microbial communities[19] often live in biofilms[20], and these biofilms are of profound importance in both medicine and microbial ecology. Research on biofilms is in its infancy; it will be many years before we more fully understand their nature and are able to use our knowledge in practical ways. In general, microbe-microbe interactions[21] have not yet been extensively explored.

(8) <u>The genomes of many microorganisms have already been sequenced, and many more will be determined in the coming years. These sequences are ideal for learning how the genome is related to cell structure[22] and what the minimum assortment of genes necessary for life is. Analysis of the genome and its activity will require continuing advances in the field of</u>

bioinformatics and the use of computers to investigate biological problems.

(9) Further research on unusual microorganisms and microbial ecology will lead to a better understanding of the interactions between microorganisms and the inanimate world[23]. Among other things, this understanding should enable us to more effectively control pollution. Similarly, it has become clear that microorganisms are essential partners with higher organisms in symbiotic relationships[24]. Greater knowledge of symbiotic relationships can help improve our appreciation of the living world. It also will lead to improvements in the health of plants, livestock, and humans.

(10) Because of their relative simplicity, microorganisms are excellent subjects for the study of a variety of fundamental questions in biology. For example, how do complex cellular structures develop and how do cells communicate with one another and respond to the environment?

(11) Finally, microbiologists will be challenged to carefully assess the implications of new discoveries and technological developments. They will need to communicate a balanced view of both the positive and negative long-term impacts of these events on society.

The future of microbiology is bright. The microbiologist René Dubos has summarized well the excitement and promise of microbiology: How extraordinary that, all over the world, microbiologists are now involved in activities as different as the study of gene structure[25], the control of disease, and the industrial processes based on the phenomenal ability of microorganisms to decompose and synthesize complex organic molecules. Microbiology is one of the most rewarding professions because it gives its practitioners the opportunity to be in contact with all the other natural sciences and thus to contribute in many different ways to the betterment of human life.

Notes to the Difficult Sentences

Microbiologists must find ways to stop the spread of established infectious diseases. Increases in antibiotic resistance will be a continuing problem, particularly the spread of multiple drug resistance that can render a pathogen impervious to current medical treatment. Microbiologists have to create new drugs and find ways to slow or prevent the spread of drug resistance. New vaccines must be developed to protect against diseases such as AIDS. It will be necessary to use techniques in molecular biology and recombinant DNA technology to solve these problems.

微生物学家必须设法阻止已经被证实的传染病的传播。微生物的抗生素抗性的增强将是一个持久的问题，特别是多药耐药性的传播，可使病原体不受当前医学治疗的影响。微生物学家不得不开发新的药物并找到减缓或阻止药物抗性传播的方法。新疫苗必须被开发出来以使人们免受像艾滋病这样的疾病的威胁。有必要使用微生物学新技术和重组 DNA 技术来解决这些问题。

The genomes of many microorganisms have already been sequenced, and many more will be determined in the coming years. These sequences are ideal for learning how the genome is related to cell structure and what the minimum assortment of genes necessary for life is. Analysis of the genome and its activity will require continuing advances in the field of bioinformatics and the use of computers to investigate biological problems.

许多微生物基因组已经被测序，并且更多微生物基因组将在未来被鉴定。这些序列非常有助于我们了解基因组与细胞结构的相互关系以及什么是生命所必需的最小基因组成。基因组及其活性的分析将需要人们在生物信息学领域不断取得进展并将计算机应用在生物学问题的研究方面。

Professional Words and Phrases

[1] **hemorrhagic** [ˈhemərædʒik] adj. 出血的
[2] **hemorrhagic fever** 出血热
[3] **antibiotic** [æntibaiˈɔtik] adj. 抗菌的
[4] **drug resistance** 药物抗性
[5] **impervious** [imˈpəːviəs] adj. 不能渗透的，不为所动的
[6] **chronic disease** 慢性病
[7] **autoimmune** [ˌɔːtouɪˈmjuːn] adj. 自身免疫的
[8] **cardiovascular disease** 心血管疾病
[9] **affliction** [əˈflikʃən] n. 痛苦，苦恼，苦难
[10] **host cell** 宿主细胞
[11] **host** [həust] n. 宿主
[12] **invasion** [inˈveiʒən] n. 侵略，侵入
[13] **environmental control** 环境控制
[14] **pollutant** [pəˈluːtənt] n. 污染物
[15] **toxic waste** 有毒废物
[16] **agricultural productivity** 农业生产率
[17] **microbial diversity** 微生物多样性
[18] **extreme environment** 极端环境
[19] **microbial community** 微生物群落
[20] **biofilm** [ˈbiɔfilm] n. 生物膜
[21] **microbe-microbe interaction** 微生物与微生物相互作用
[22] **cell structure** 细胞结构
[23] **inanimate world** 非生命界
[24] **symbiotic relationship** 共生作用
[25] **gene structure** 基因结构

Exercises

1. Matching

1) hemorrhagic fever   a) non-living world

2) drug resistance   b) an aggregate of microorganisms in which cells adhere to each other or a surface

3) autoimmune   c) the reduction in effectiveness of a drug such as an antimicrobial or an antineoplastic in curing a disease or condition

4) extreme environment   d) a relationship between two entities which is mutually beneficial for the participants of the relationship

5) biofilm   e) of, relating to, or caused by autoantibodies or T cells that attack

6) symbiotic relationship   molecules, cells, or tissues of the organism producing them
7) inanimate world

6) symbiotic relationship
7) inanimate world
   f) extreme conditions which are challenging to most life forms
   g) a diverse group of animal and human illnesses that are caused by four distinct families of RNA viruses

2. True or False
1) Microbial diversity is needed for considerable research.
2) We have known the nature of microbial communities that live in biofilms.
3) The sequencing and analysis of the genomes of many microorganisms have already been completed.
4) The antibiotic resistance of a pathogen to multiple drugs is a threat to stopping the spread of established infectious diseases.
5) Microbiological research is needed on the association between infectious agents and chronic diseases.
6) Microorganisms are essential partners with higher organisms in symbiotic relationships.
7) We have clearly understood how pathogens interact with host cells and the ways in which diseases arise.

3. Reading Comprehension
1) Which is not an example of new and reemerging infectious diseases?
  A. AIDS
  B. hemorrhagic fevers
  C. tuberculosis
  D. None of the above
2) Which of the following statements is not true?
  A. New vaccines have been developed to protect against diseases such as AIDS.
  B. The symbiotic relationships can help improve our appreciation of the living world.
  C. The microbe-microbe interactions have not yet been extensively explored.
  D. The discovery of new microorganisms may well lead to further advances in industrial processes and enhanced environmental control.
3) What can microorganisms help us do?
  A. To serve as sources of high-quality food and other practical products such as enzymes for industrial applications
  B. To degrade pollutants and toxic wastes
  C. To be used as vectors to treat diseases and enhance agricultural productivity
  D. All

4. Translation from English to Chinese
Microorganisms are increasingly important in industry and environmental control, and we must learn how to use them in a variety of new ways. For example, microorganisms can (a) serve as sources of high-quality food and other practical products such as enzymes for industrial applications, (b) degrade pollutants and toxic wastes, and (c) be used as vectors to treat diseases and enhance agricultural productivity. There also is a continuing need to protect food and crops from microbial damage.

5. Translation from Chinese to English

进一步研究特殊微生物以及微生物生态学将使得我们更好地理解微生物与非生命界的相互关系。此外，理解这一相互关系能使我们更有效地控制污染。类似的是，微生物是其共生的高等生物所必需的已经是非常清楚的。更多地了解这一共生关系将有助于我们对生命界的深入认识以及植物、家畜和人的健康的改善。

2.3 Prokaryotes, Eukaryotic Microbes and Viruses

The prokaryotes[1] are a group of organisms that lack a cell nucleus[2], or any other membrane-bound organelles[3]. They differ from the eukaryotes[4], which have a cell nucleus. Most are unicellular[5], but a few prokaryotes such as myxobacteria[6] have multicellular[7] stages in their life cycles. It is also spelled "procaryote".

The prokaryotes are divided into two domains: the bacteria and the archaea. Archaea were recognized as a domain of life in 1990. These organisms were originally thought to live only in inhospitable conditions such as extremes[8] of temperature, pH, and radiation but have since been found in all types of habitats.

The classification of eukaryotic[9] microbes[10] (also called eucaryotic microbes) is problematic and has changed frequently. Historical schemes based on similarity in morphology and chemistry have been replaced with schemes based on nucleotide sequences[11] and ultrastructural[12] features. There are three major eukaryotic microorganisms: fungi, algae and protozoa.

Microbiologists use the term fungus [pl., fungi] to include eucaryotic, spore-bearing[13] organisms with absorptive[14] nutrition, no chlorophyll[15], and that reproduce[16] sexually[17] and asexually[18]. Scientists who study fungi are mycologists[19], and the scientific discipline dealing with fungi is called mycology[20]. The study of fungal toxins[21] and their effects is called mycotoxicology[22], and the diseases caused by fungi in animals are known as mycoses[23] [s., mycosis]. The five-kingdom system places the fungi in the kingdom Fungi[24]. According to the universal phylogenetic[25] tree[26], fungi are members of the domain Eucarya[27]. As presently delimitated, the kingdom Fungi is believed to constitute a monophyletic[28] group referred to as the true fungi or Eumycota[29].

Phycology or algology is the study of algae. The word phycology is derived from the Greek phykos, meaning seaweed. The term algae [s., alga] was originally used to define simple "aquatic[30] plants[31]". As noted above, it no longer has any formal significance in classification schemes[32]. Instead the algae can be described as eucaryotic organisms that have chlorophyll a[33] and carry out oxygen-producing photosynthesis[34]. They differ from other photosynthetic eucaryotes in lacking a well-organized vascular[35] conducting system[36] and in having very simple reproductive structures. In sexual reproduction[37] the whole organism may serve as a gamete[38]; unicellular structures (gametangia[39]) may produce gametes; or gametes can be formed by multicellular gametangia in which every cell is fertile. Unlike the case with plants, algal[40] gametangia do not have nonfertile[41] cells.

The microorganisms called protozoa are studied in the discipline called protozoology[42]. A protozoan can be defined as a usually motile eukaryotic unicellular protist[43]. Protozoa are directly related only on the basis of a single negative characteristic—they are not multicellular. All, however, demonstrate the basic body plan of a single protistan[44] eucaryotic cell. Most protozoa are free living and inhabit freshwater or marine environments. Many terrestrial[45] protozoa can be found in decaying organic matter, in soil, and even in beach sand; some are parasitic[46] in plants or animals.

Viruses are infectious agents with fairly simple, acellular[47] organization. They possess only one type of nucleic acid[48], either DNA or RNA, and only reproduce within living cells. Clearly viruses are quite different from procaryotic and eucaryotic microorganisms, and are studied by virologists. Despite their simplicity in comparison with cellular organisms, viruses are extremely important and deserving of close attention. The study of viruses has contributed significantly to the discipline of molecular biology. Many human viral diseases are already known and more are discovered or arise every year, as demonstrated by the appearance of AIDS. The whole field of genetic engineering is based in large part upon discoveries in virology (the study of viruses). Thus it is easy to understand why virology is such a significant part of microbiology.

<u>Viruses are a unique group of infectious agents whose distinctiveness resides in their simple, acellular organization and pattern of reproduction. A complete virus particle or virion consists of one or more molecules of DNA or RNA enclosed in a coat of protein, and sometimes also in other layers. These additional layers may be very complex and contain carbohydrates[49], lipids[50], and additional proteins[51]. Viruses can exist in two phases: extracellular[52] and intracellular[53]. Virions[54], the extracellular phase, possess few if any enzymes and cannot reproduce independent of living cells. In the intracellular phase, viruses exist primarily as replicating nucleic acids that induce host metabolism to synthesize virion components; eventually complete virus particles[55] or virions are released.</u>

Notes to the Difficult Sentences

Microbiologists use the term fungus [pl., fungi] to include eucaryotic, spore-bearing organisms with absorptive nutrition, no chlorophyll, and that reproduce sexually and asexually. Scientists who study fungi are mycologists, and the scientific discipline dealing with fungi is called mycology. The study of fungal toxins and their effects is called mycotoxicology, and the diseases caused by fungi in animals are known as mycoses [s., mycosis]. The five-kingdom system places the fungi in the kingdom Fungi. According to the universal phylogenetic tree, fungi are members of the domain Eucarya. As presently delimitated, the kingdom Fungi is believed to constitute a monophyletic group referred to as the true fungi or Eumycota.

微生物学家使用专业术语"真菌"来表示真核的、产生孢子的、需要吸收营养的、没有叶绿素的、进行有性和无性生殖的生物。研究真菌的科学家是真菌学家，而研究真菌的科学则被称为真菌学。研究真菌毒素及其作用的科学称为真菌毒理学，而由真菌导致的动物病症称为霉菌病。五界系统学说将真菌归为真菌界。根据通用系统发育树，真菌属于真核生物域。正像当今生物定界的标准所认为的，真菌界指的是真正的真菌或真菌门的一个单一的群体。

Viruses are a unique group of infectious agents whose distinctiveness resides in their simple, acellular organization and pattern of reproduction. A complete virus particle or virion consists of one or more molecules of DNA or RNA enclosed in a coat of protein, and sometimes also in other layers. These additional layers may be very complex and contain carbohydrates, lipids, and additional proteins. Viruses can exist in two phases: extracellular and intracellular. Virions, the extracellular phase, possess few of any enzymes and cannot reproduce independent of living cells. In the intracellular phase, viruses exist primarily as replicating nucleic acids that induce host metabolism to synthesize virion components; eventually complete virus particles or virions are released.

　　病毒是一类独一无二的病原体，它的独特之处在于简单的非细胞组成及其繁殖方式。完整的病毒颗粒是由蛋白质外壳和包裹在其中的一条或多条 DNA 或 RNA 组成的，但蛋白质外壳有时也在其他外层结构里。这一辅助的外层结构有时也非常复杂，可能是由碳水化合物、脂和其他的蛋白质组成。病毒存在两个阶段：细胞内阶段和细胞外阶段。病毒颗粒是细胞外阶段的形式，在此阶段病毒几乎没有什么酶并且不能独立于活细胞繁殖。在细胞内阶段，病毒主要以复制的核酸形式存在，这些核酸能诱导宿主代谢合成病毒颗粒的组成成分，最后导致宿主释放完整的病毒颗粒。

Professional Words and Phrases

[1] **prokaryote** [prəuˈkæriɔt] n. 原核生物
[2] **nucleus** [ˈnjuːkliəs] n. 细胞核
[3] **organelle** [ɔːgəˈnel] n. 细胞器
[4] **eukaryote** [juˈkæriəut] n. 真核生物
[5] **unicellular** [ˌjuːniˈseljulə] adj. 单细胞的
[6] **myxobacteria** [miksəbækˈtiəriə] n. 黏细菌
[7] **multicellular** [mʌltiˈseljulə] adj. 多细胞的
[8] **extreme** [iksˈtriːm] adj. 极度的，极端的；n. 极端，极限
[9] **eukaryotic** [juˈkæriəutik] adj. 真核细胞的
[10] **eukaryotic microbes** 真核微生物
[11] **nucleotide sequence** 核苷酸序列
[12] **ultrastructural** [ˈʌltrəlˈstrʌktʃərəl] adj. 超微结构的
[13] **spore-bearing** adj. 产生孢子的
[14] **absorptive** [əbˈsɔːptiv] adj. 吸收性的，有吸收力的
[15] **chlorophyll** [ˈklɔrəfil] n. 叶绿素
[16] **reproduce** [riːprəˈdjuːs] v. 再生，复制，生殖
[17] **sexually** [ˈsekʃuəli] adv. 有性地
[18] **asexually** [eiˈsekʃuəli] adv. 无性地
[19] **mycologist** [maiˈkɔlədʒist] n. 真菌学家
[20] **mycology** [maiˈkɔlədʒi] n. 真菌学
[21] **fungal toxin** 真菌毒素
[22] **mycotoxicology** [maikəutɔksəˈkɔlədʒi] n. 真菌毒理学
[23] **mycoses** [maiˈkəusis] n. 霉菌病，mycosis 的复数形式
[24] **kingdom Fungi** 真菌界

[25] **phylogenetic**   [failəudʒə'netik]   adj. 系统发生的
[26] **phylogenetic tree**   系统发育树
[27] **Eucarya**   真核生物域
[28] **monophyletic**   [mɔnəufai'letik]   adj. 单元的，单源的
[29] **Eumycota**   ['jumaikəutə]   n. 真菌门
[30] **aquatic**   [ə'kwætik]   adj. 水生的
[31] **aquatic plant**   水生植物
[32] **classification scheme**   分类表
[33] **chlorophyll a**   叶绿素 a
[34] **photosynthesis**   [fəutəu'sinθəsis]   n. 光合作用
[35] **vascular**   ['væskjulə]   adj. 脉管的
[36] **vascular conducting system**   维管系统
[37] **sexual reproduction**   有性生殖
[38] **gamete**   ['gæmi:t]   n. 配子
[39] **gametangia**   ['gæmə'tændʒiəm]   n. 配子囊
[40] **algal**   ['ælgəl]   adj. 藻类的
[41] **nonfertile**   ['nʌn'fə:tail]   adj. 不育的，不孕的
[42] **protozoology**   [prəutəzəu'ɔlədʒi]   n. 原生动物学
[43] **protist**   ['prəutist]   n. 原生生物
[44] **protistan**   [prəu'tistən]   adj. 原生生物的
[45] **terrestrial**   [ti'restriəl]   adj. 陆地的，陆生的
[46] **parasitic**   [pærə'sitik]   adj. 寄生的
[47] **acellular**   [ei'seljulə]   adj. 非细胞的，非细胞组成的
[48] **nucleic acid**   核酸
[49] **carbohydrate**   ['kɑ:bəu'haidreit]   n. 碳水化合物
[50] **lipid**   ['lipid]   n. 脂
[51] **protein**   ['prəuti:n]   n. 蛋白质
[52] **extracellular**   [ekstrə'seljulə]   adj. (位于或发生于)细胞外的
[53] **intracellular**   [intrə'seljulə]   adj. 细胞内的
[54] **virion**   ['vaiəriɔn]   n. 病毒粒子
[55] **virus particle**   病毒粒子

Exercises

1. Matching

1) prokaryote   a) virus particle
2) eukaryote   b) slime bacteria
3) myxobacteria   c) biological molecules essential for life, and include DNA
                                (deoxyribonucleic acid) and RNA (ribonucleic acid)
4) eukaryotic microbes   d) a group of organisms that have a cell nucleus
5) chlorophyll   e) a group of organisms that lack a cell nucleus, or any other
                                membrane-bound organelles

6) gamete
7) nucleotide
8) nucleic acid
9) virion

f) microorganisms that have a cell nucleus
g) a green pigment found in almost all plants, algae, and cyanobacteria
h) molecules that, when joined together, make up the structural units of RNA and DNA
i) a cell that fuses with another cell during fertilization in organisms that reproduce sexually

2. True or False

1) The prokaryotes lack a cell nucleus, but some of them have membrane-bound organelles.
2) The classification of eukaryotic microbes has currently been based on similarity in morphology and chemistry.
3) Fungi are thought to be the microorganisms, which include eucaryotic, spore-bearing organisms with absorptive nutrition, no chlorophyll, and that reproduce sexually and asexually.
4) Virologists commonly agree that viruses are procaryotic microbes, which are quite different from eucaryotic microorganisms.
5) A complete virus particle or virion consists of one or more molecules of DNA or RNA enclosed in a coat of protein, and sometimes also in other layers.

3. Reading Comprehension

1) Which are not included in the eukaryotic microorganisms?
   A. fungi
   B. algae
   C. protozoa
   D. bacteria

2) Which of the statements about algae below is not true?
   A. having chlorophyll a
   B. having a well-organized vascular conducting system
   C. carrying out oxygen-producing photosynthesis
   D. having very simple reproductive structures

3) What is not the characteristic of protozoa?
   A. motile
   B. eukaryotic
   C. multicellular
   D. Most inhabit freshwater or marine environments

4. Translation from English to Chinese

Clearly viruses are quite different from procaryotic and eucaryotic microorganisms, and are studied by virologists. Despite their simplicity in comparison with cellular organisms, viruses are extremely important and deserving of close attention. The study of viruses has contributed significantly to the discipline of molecular biology. Many human viral diseases are already known and more are discovered or arise every year.

5. Translation from Chinese to English

真核微生物的分类仍存在很大争议并且是经常变化的。以前的分类方案依据的是形态和化学的相似性，但现在的分类方案依据的是核酸序列及超微结构的相似性。目前主要有三类主要的真核微生物：真菌、藻类和原生动物。

2.4 Extreme Microbes—Additional Reading

One of the first extreme microbes to make a big mark on science was *Thermus aquaticus*[1], a bacterium first isolated from hot geysers[2] in Yellowstone National Park[3]. The bacterium can survive and reproduce at 66℃, a temperature ideal for copying DNA in the laboratory (heat speeds up reactions so they can complete faster). As a result, enterprising scientists stole its key DNA-copying protein, *Taq* polymerase[4], for laboratory use. Today, *Taq* is used in most biological labs around the world and is one of the reasons that forensic DNA analysis[5] is possible.

Taq, however, is not the hottest of the hot. In 2003, 200 miles from Puget Sound, scientists discovered a microbe they call "strain 121" that can survive at temperatures higher than water's boiling point—they can, in fact, reproduce at temperatures as high as 121℃ (hence their name). These are archaea, single-celled organisms similar to bacteria, and are the most heat-tolerant[6] microbes ever discovered. Interestingly, strain 121 breathes not oxygen but iron oxide.

Then there's the other extreme: cold. After a grueling search for life in the Antarctic "ice-block" lake known as Lake Vida -which, under a thick layer of ice, harbors salty water as cold as −10℃—scientists uncovered, in 2002, microbes at least 2,800 years old. "It was some very cold drilling," scientist Peter Doran of the University of Illinois at Chicago told the National Science Foundation. "We were there for two weeks at temperatures approaching −40℃." These microbes might be very similar to those that could be found on Mars[7], he added -relevant given that the Mars Phoenix Lander will be searching for signs of life in the Martian arctic this May. "Mars is believed to have a water-rich past, and if life developed, a Lake Vida-type ecosystem may have been the final niche[8] for life on Mars before the water bodies froze solid," Doran said.

Other microbes might be able to keep us from aging as quickly. <u>A species of bacteria called *Deinococcus radiodurans*[9] was discovered in Oregon in 1956 when scientists who were attempting to sterilize[10] food with high doses of radiation found that even extremely high doses[11] could not kill these bacteria. Radiation contains so much energy that it can kill animals quickly at high enough doses; even at low levels, exposure is dangerous because it damages DNA and causes cancer. Scientists are therefore studying these tough microbes in the hopes of understanding why they are so good at repairing their own DNA—and to learn whether we can co-opt their skills to prevent both cancer and aging.</u>

Other microbes were recently discovered that actually eat the byproducts of radiation. Scientists found a species of bacteria in South Africa living two miles (1mile=1609 meters) below Earth's surface in high-pressure environments. These bacteria, existing without sun and all other forms of life, and which may have been lurking in those depths for up to 25 million years, get their nourishment from radiation byproducts produced by the reaction of uranium[12], thorium[13] and potassium[14] with water over time. They might be the oldest ancestors of current bacteria, raising the question of whether life actually began underground.

Our planet clearly harbors a diverse group of microbes living in the harshest environments imaginable-and we've probably only seen the tip of the iceberg, so to speak. As scientists continue to uncover life-forms adapted to the darkest and deadliest places, they will not only learn a thing or two about survival, but might also get a sneak peek of possible life on other planets.

Notes to the Difficult Sentences

A species of bacteria called *Deinococcus radiodurans* was discovered in Oregon in 1956 when scientists who were attempting to sterilize food with high doses of radiation found that even extremely high doses could not kill these bacteria. Radiation contains so much energy that it can kill animals quickly at high enough doses; even at low levels, exposure is dangerous because it damages DNA and causes cancer. Scientists are therefore studying these tough microbes in the hopes of understanding why they are so good at repairing their own DNA—and to learn whether we can co-opt their skills to prevent both cancer and aging.

一种被称为耐辐射球菌的细菌是在 1956 年美国的俄勒冈州被发现的，科学家当时试图通过高剂量辐射方法对食品进行灭菌时发现，即使高剂量辐射也无法杀死这类细菌。辐射剂量高到一定程度时足以杀死动物；即使暴露在低水平的辐射下也是危险的，因为辐射能损害 DNA 并导致癌症。因此，科学家正在研究这类极端微生物，希望弄清楚为什么它们如此擅长修复自身的 DNA，并且想知道我们是否能利用这些技术阻止癌症和衰老的发生。

Professional Words and Phrases

[1] *Thermus aquaticus*   水生栖热菌
[2] geyser   ['gaizə]   n. 间歇喷泉
[3] Yellowstone National Park   （美国）黄石国家公园
[4] polymerase   ['pɔliməreis]   n. 聚合酶
[5] DNA analysis   （法医的）DNA 分析
[6] heat-tolerant   adj. 耐热的
[7] Mars   [mɑːz]   n. 火星
[8] niche   [nitʃ]   n. 生境
[9] *Deinococcus radiodurans*   耐辐射球菌
[10] sterilize   ['sterilaiz]   vt. 使不育，杀菌
[11] dose   [dəus]   n. 剂量
[12] uranium   [juə'reiniəm]   n. 铀
[13] thorium   ['θɔːriəm]   n. 钍
[14] potassium   [pə'tæsiəm]   n. 钾

Exercises

1. Matching

1) *Taq* polymerase   a) the chemical element with the symbol K
2) DNA analysis   b) a naturally occurring radioactive chemical element with the symbol Th
3) niche   c) a silvery-white metallic chemical element with the symbol U
4) sterilize   d) a thermostable DNA polymerase named after the thermophilic bacterium *Thermus aquaticus*

5) uranium e) any technique used to analyze genes and DNA
6) thorium f) a term referring to any process that eliminates (removes) or kills all forms of life, including transmissible agents
7) potassium g) a habitat supplying the factors necessary for the existence of an organism or species

2. True or False

1) *Thermus aquaticus* can survive and reproduce at 121℃.
2) The strain 121 is a kind of extreme microbe that can tolerate the cold.
3) *Deinococcus radiodurans* might be able to keep us from aging as quickly.
4) Some microbes were recently discovered to eat the byproducts of radiation.
5) The animals can survive at high enough doses of radiation like extreme microbes.

3. Reading Comprehension

1) Which does not belong to the extreme microbes?
   A. heat-tolerant microbes
   B. cold-tolerant microbes
   C. radiation-tolerant microbes
   D. None of the above

2) Which statement is not correct about *Taq* polymerase?
   A. *Taq* polymerase is a key DNA-copying protein.
   B. *Taq* polymerase can tolerate 121℃.
   C. *Taq* polymerase makes the forensic DNA analysis possible.
   D. *Taq* polymerase is used widely in the biological labs.

3) Which is not the reason of why the scientists study *Deinococcus radiodurans*?
   A. Because *Deinococcus radiodurans* might be able to keep us from aging as quickly.
   B. To understand why they are so good at repairing their own DNA.
   C. To kill the bacteria.
   D. To utilize the skills that *Deinococcus radiodurans* tolerate the radiation.

4. Translation from English to Chinese

One of the first extreme microbes to make a big mark on science was *Thermus aquaticus*, a bacterium first isolated from hot geysers in Yellowstone National Park. The bacterium can survive and reproduce at 66℃, a temperature ideal for copying DNA in the laboratory (heat speeds up reactions so they can complete faster). As a result, enterprising scientists stole its key DNA-copying protein, *Taq* polymerase, for laboratory use. Today, *Taq* is used in most biological labs around the world and is one of the reasons that forensic DNA analysis is possible.

5. Translation from Chinese to English

最近，人们发现有些微生物能利用辐射的副产物。科学家在南非境内距离地表2英里（1英里=1609米）处的这一高压力环境中发现了一类细菌。这类细菌是在没有阳光和其他生命的地方存在的，并且在此深度可能存在了2500万年，它们是从铀、钍和钾与水反应产生的副产物中获取营养的。它们可能是当今细菌最早的祖先，这让我们怀疑生命是否始于地下。

Chapter 3   Cellular Biology

[本章中文导读]

　　细胞作为生命体最基本的结构单位,从被 Robert Hooke 发现直至细胞能够被独立培养以来,与细胞有关的生物学知识逐步发展成为生物学的一个重要分支学科,并且对当今生物学的各个分支学科的发展起到不可估量的作用。本章主要介绍了细胞的发现(第3.1 节)、细胞的基本属性(第 3.2 节)以及哺乳动物细胞培养的概念(第 3.3 节)等细胞生物学方面的一些基础专业知识。作者想通过这些专业知识的英文讲解,增加学生对细胞生物学有关知识学习的兴趣,达到使学生初步掌握一些在细胞生物学领域常见的专业英语词汇的具体含义和实际用法的目的。

3.1　The Discovery of Cells

　　We do not know when humans first discovered the remarkable ability of curved-glass surfaces to bend light and form images. Spectacles were first made in Europe in the thirteenth century, and the first compound (double-lensed) microscopes[1] were constructed by the end of the sixteenth century. By the mid-1600s, a handful of pioneering scientists had used their handmade microscopes to uncover a world that would never have been revealed to the naked eye. The discovery of cells is generally credited to Robert Hooke, an English microscopist[2] who, at age 27, was awarded the position of curator of the Royal Society[3], England's foremost scientific academy. One of the many questions Hooke attempted to answer was why stoppers made of cork (part of the bark of trees) were so well suited to holding air in a bottle. As he wrote in 1665: "I took a good clear piece of cork, and with a pen-knife sharpen'd as keen as a razor, I cut a piece of it off and ... then examining it with a microscope, me thought I could perceive it to appear a little porous ... much like a honeycomb." Hooke called the pores cells because they reminded him of the cells inhabited by monks living in a monastery. In actual fact, Hooke had observed the empty cell walls[4] of dead plant tissue, walls that had originally been produced by the living cells they surrounded.

　　Meanwhile, Anton van Leeuwenhoek, a Dutchman who earned a living selling clothes and buttons, was spending his spare time grinding lenses and constructing microscopes of remarkable quality. For 50 years, Leeuwenhoek sent letters to the Royal Society of London describing his microscopic[5] observations-along with a rambling discourse on his daily habits and the state of his health. Leeuwenhoek was the first to examine a drop of pond water under the microscope and, to his amazement, to observe the teeming microscopic "animalcules[6]" that darted back and forth before his eyes. He was also the first to describe various forms of bacteria, which he obtained from water in which pepper had been soaked and from scrapings of his teeth. His initial letters to the Royal Society describing this previously unseen world were met with such skepticism that the

Society dispatched its curator, Robert Hooke, to confirm the observations. Hooke did just that, and Leeuwenhoek was soon a worldwide celebrity, receiving visits in Holland from Peter the Great of Russia and the queen of England.

It wasn't until the 1830s that the widespread importance of cells was realized. In 1838, Matttuas Schleiden, a German lawyer turned botanist[7], concluded that, despite differences in the structure of various tissues, plants were made of cells and that the plant embryo[8] arose from a single cell. In 1839, Theodor Schwann, a German zoologist[9] and colleague of Schleiden's, published a comprehensive report on the cellular[10] basis of animal life. Schwann concluded that the cells of plants and animals are similar structures and proposed these two tenets of cell theory: ①All organisms are composed of one or more cells. ②The cell is the structural unit of life.

Schleiden and Schwann's ideas on the *origin of cells* proved to be less insightful; both agreed that cells could arise from noncellular[11] materials. Given the prominence that these two scientists held in the scientific world, it took a number of years before observations by other biologists were accepted as demonstrating that cells did not arise in this manner any more than organisms arose by spontaneous generation. By 1855, Rudolf Virchow, a German pathologist[12], had made a convincing case for the third tenet of the cell theory: Cells can arise only by division[13] from a preexisting cell.

Notes to the Difficult Sentences

For 50 years, Leeuwenhoek sent letters to the Royal Society of London describing his microscopic observations -along with a rambling discourse on his daily habits and the state of his health. Leeuwenhoek was the first to examine a drop of pond water under the microscope and, to his amazement, to observe the teeming microscopic "animalcules" that darted back and forth before his eyes. He was also the first to describe various forms of bacteria, which he obtained from water in which pepper had been soaked and from scrapings of his teeth. His initial letters to the Royal Society describing this previously unseen world were met with such skepticism that the society dispatched its curator, Robert Hooke, to confirm the observations.

列文虎克（Leeuwenhoek）在50年里写给伦敦皇家学会的很多信件中描述了他在显微镜下观察的东西，以及他的日常生活习惯和健康情况等琐碎话题。列文虎克是首次使用显微镜观察池塘的一滴水的人。令他非常吃惊的是，他观察到了一些"微小的动物"在眼前快速地来回游动。他也首次描述了各种细菌的形状，这些细菌是他从浸泡过的辣椒和他牙屑中获得的。起初他给皇家学会的信中描述的未知世界备受质疑，以至于皇家学会指派学会的管理者Robert Hooke对这些发现进行了验证。

Professional Words and Phrases

[1] **microscope** ['maikrəskəup] n. 显微镜
[2] **microscopist** [mai'krɔskəpist] n. 使用显微镜的技术人员
[3] **the Royal Society** （英国）皇家学会
[4] **cell wall** 细胞壁
[5] **microscopic** ['maikrə'skɔpik] adj. 显微镜的，微观的
[6] **animalcule** ['æniməl'kjul] n. 微小动物

[7] **botanist** ['bɔtənist] n. 植物学家
[8] **embryo** ['embriəu] n. 胚胎
[9] **zoologist** [zəu'ɔlədʒist] n. 动物学家
[10] **cellular** ['seljulə] adj. 细胞的
[11] **noncellular** [nɔn'seljulə] adj. 非细胞的
[12] **pathologist** [pæ'θɔlədʒist] n. 病理学家
[13] **division** [di'viʒən] n. 分裂，分开

Exercises

1. Matching
1) microscope   a) without a cell structure
2) cell wall   b) a scientist who studies plants
3) animalcule   c) a doctor who studies the cause and the development of a disease
4) botanist   d) the tough, usually flexible but sometimes fairly rigid layer that surrounds some types of cells
5) noncellular   e) an instrument used to see objects too small for the naked eye
6) zoologist   f) an older term for a microscopic animal or protozoan
7) pathologist   g) a biologist who studies animals

2. True or False
1) The discovery of cells is generally credited to Robert Hooke.
2) Robert Hooke was the first to describe various forms of bacteria.
3) The cells could arise from noncellular materials.

3. Reading Comprehension
1) Who made the important contributions to the discovery of cells?
　A. Robert Hooke
　B. Anton van Leeuwenhoek
　C. Matttuas Schleiden
　D. All
2) Which is not included in the cell theory?
　A. The cells could arise from noncellular materials.
　B. The cell is the structural unit of life.
　C. Cells can arise only by division from a preexisting cell.
　D. All organisms are composed of one or more cells.
3) Which of the following statements is not true?
　A. The first compound (double-lensed) microscopes were constructed by the end of the sixteenth century.
　B. Rudolf Virchow demonstrated that cells can arise only by division from a preexisting cell.
　C. Robert Hooke observed the real living cells.
　D. Schleiden and Schwann's ideas on the *origin of cells* proved to be not right.

4. Translation from English to Chinese
　　It wasn't until the 1830s that the widespread importance of cells was realized. In 1838, Matttuas Schleiden, a German lawyer turned botanist, concluded that, despite differences in

the structure of various tissues, plants were made of cells and that the plant embryo arose from a single cell. In 1839, Theodor Schwann, a German zoologist and colleague of Schleiden's, published a comprehensive report on the cellular basis of animal life. Schwann concluded that the cells of plants and animals are similar structures and proposed these two tenets of cell theory: ①All organisms are composed of one or more cells. ②The cell is the structural unit of life.

5. Translation from Chinese to English

Schleiden 和 Schwann 对细胞起源的观点认为细胞来自非细胞物质，对后来的科学研究产生了不利的影响。然而，由于两位科学家在科学界享有崇高的威望，其他科学家证明细胞并不是以这种方式产生并且生命体不是通过自然发生产生的观点很多年后才被接受。直到 1855 年，一个德国的病理学家 Rudolf Virchow 提出了令人信服的佐证才完善了细胞理论的第三个原则，即细胞只能通过分裂从已存在的细胞产生。

3.2 Basic Properties of Cells

Just as plants and animals are alive, so too are cells. Life, in fact, is the most basic property of cells, and cells are the smallest units to exhibit this property. Unlike the parts of a cell, which simply deteriorate if isolated[1], whole cells can be removed from a plant or animal and cultured[2] in a laboratory where they will grow and reproduce for extended periods of time. The first culture of human cells was begun by George Gey of Johns Hopkins University in 1951. The cells were obtained from a malignant tumor and named HeLa cells after the donor, Henrietta Lacks. HeLa cells—descended by cell division[3] from this first cell sample[4]—are still being grown in laboratories around the word today. Because they are so much simpler to study than cells situated within the body, cells grown in vitro[5] (i.e., in culture, outside the body) have become an essential tool of cell and molecular biologists.

1) Cells Are Highly Complex and Organized.

Complexity is a property that is evident but difficult to describe. For the present, we can think of complexity in terms of order and consistency. The more complex a structure, the greater the number of parts that must be in their proper place, the less tolerance of errors in the nature and interactions of the parts, and the more regulation or control that must be exerted to maintain the system. We will have occasion to consider the complexity of life at several different level. We will discuss the organization of atoms into small-sized molecules; the organization of these molecules into giant polymers; and the organization of different types of polymeric molecules into complexes, which in turn are organized into subcellular[6] organelles and finally into cells. As will be apparent, there is a great deal of consistency at every level. Each type of cell has a consistent appearance in the electron microscope[7]; that is, its organelles have a particular shape and location, from one individual of a species to another. Similarly, each type of organelle has a consistent composition of macromolecules[8], which are arranged in a predictable pattern.

2) Cells Possesses a Genetic Program and the Means to Use it.

Organisms are built according to information encoded in a collection of genes. The human genetic program[9] contains enough information, if converted to words, to fill millions of pages of text. Remarkably, this vast amount of information is packaged into a set of chromosomes[10] that occupy the space of a cell nucleus[11] -thousands of times smaller than the dot on this *i*.

Genes are more than storage lockers for information: they constitute the blueprints for constructing cellular structures, the directions for running cellular activities, and the program for making more of themselves. Discovering the mechanisms by which cells use their genetic information to accomplish these functions has been one of the greatest achievements of science in recent decades.

3) Cells Are Capable of Producing More of Themselves.

Just as individual organisms are generated by reproduction, so too are individual cells. Cells reproduce by division, a process in which the contents of a "mother" cell are distributed into two "daughter" cells. Prior to division, the genetic material[12] is faithfully duplicated, and each daughter cell receives a complete and equal share of genetic information[13]. In most cases, the two daughter cells produced by division have approximately equal volume. In some cases, however, as occurs when a human oocyte[14] undergoes division, one of the cells can retain nearly all of the cytoplasm, even though it receives only half of the genetic material.

4) Cells Acquire and Utilize Energy.

Developing and maintaining complexity requires the constant input of energy. Virtually all of the energy required by life on the Earth's surface arrives in the form of electromagnetic radiation[15] from the sun. The energy of light is trapped by light-absorbing pigments[16] present in the membranes of photosynthetic cells. Light energy is converted by photosynthesis into chemical energy that is stored in energy-rich carbohydrates, such as sucrose[17] or starch[18]. For most animal cells, energy arrives prepackaged, often in the form of the sugar glucose[19]. In humans, glucose is released by the liver into the blood where it circulates through the body delivering chemical energy[20] to all the cells. Once in a cell, the glucose is disassembled in such a way that its energy content can be stored in a readily available form (usually as ATP) and later put to use in running all of the cell's myriad energy-requiring activities.

5) Cells Carry Out a Variety of Chemical Reactions.

Cells function like miniaturized chemical plants. Even the simplest bacterial cell is capable of hundreds of different chemical transformations, none of which occurs at any significant rate in the inanimate world. Virtually all chemical changes that take place in cells require enzymes—molecules that greatly increase the rate at which a chemical reaction[21] occurs. The sum total of the chemical reactions in a cell represents that cell's metabolism.

6) Cells Engage in Numerous Mechanical Activities.

Cells are sites of bustling activity. Materials are transported from place to place, structures are assembled and then rapidly disassembled, and, in many cases, the entire cell moves itself from one site to another. These types of activities are based on dynamic, mechanical changes within cells, most of which are initiated by changes in the shape of certain "motor" proteins.

7) Cells Are Able to Respond to Stimuli.

Some cells respond to stimuli in obvious ways; a single-celled protist, for example, moves away from an object in its path or moves toward a source of nutrients. Cells within a multicellular[22] plant or animal respond to stimuli less obviously, but they respond, nonetheless.

Most cells are covered with receptors[23] that interact with substances in the environment in highly specific ways. Cells possess receptors to hormones, growth factors[24], extracellular materials, as well as to substances on the surfaces of other cells. A cell's receptors provide doorways through which external agents can evoke specific responses in target cells. Cells may respond to specific stimuli by altering their metabolic activities[25], preparing for cell division, moving from one place to another, or even committing suicide.

8) Cells Are Capable of Self-Regulation.

In addition to requiring energy, maintaining a complex, ordered state requires constant regulation. The importance of a cell's regulatory mechanisms[26] becomes most evident when they break down. For example, failure of a cell to correct a mistake when it duplicates its DNA may result in a debilitating mutation[27], or a breakdown in a cell's growth control can transform the cell into a cancer cell with the capability of destroying the entire organism. We are gradually learning more and more about how the cell controls its activities, but much more is left to discover. In the cell, the information for product design resides in the nucleic acids, and the construction workers are primarily proteins. It is the presence of these two types of macromolecules that, more than any other factor, sets the chemistry of the cell apart from that of the nonliving world. In the cell, the workers must act without the benefit of conscious direction. Each step of a process must occur spontaneously in such a way that the next step is automatically triggered. All the information required to direct a particular activity—whether it be the synthesis of a protein, the secretion[28] of a hormone, or the contraction[29] of a muscle fiber[30]—must be present within the system itself. In many ways, cells operate in a manner analogous to the orange-squeezing contraption discovered by "The Professor". Each type of cellular activity requires a unique set of molecular tools and machines.

Notes to the Difficult Sentences

Complexity is a property that is evident but difficult to describe. For the present, we can think of complexity in terms of order and consistency. The more complex a structure, the greater the number of parts that must be in their proper place, the less tolerance of errors in the nature and interactions of the parts, and the more regulation or control that must be exerted to maintain the system. We will have occasion to consider the complexity of life at several different level. We will discuss the organization of atoms into small-sized molecules; the organization of these molecules into giant polymers; and the organization of different types of polymeric molecules into complexes, which in turn are organized into subcellular organelles and finally into cells.

复杂性是细胞显著的特征，但这一特征又是很难描述的。目前我们使用有序性和一致性来描述这一复杂性。结构越复杂，处于有序状态的各组成部分的数量就越大，组成部分的相互作用以及自身耐受错误的能力就越小，用于维持系统调控的机制就越多。我们可以从不同水平上考虑生命的复杂性。我们将讨论原子组成小分子，分子又组成巨大的聚合物，不同的聚合物分子又组成复合物，进而这些复合物又组成亚细胞结构的细胞器并最终组成细胞。

Cells function like miniaturized chemical plants. Even the simplest bacterial cell is capable of hundreds of different chemical transformations, none of which occurs at any significant rate in the inanimate world. Virtually all chemical changes that take place in cells require enzymes—molecules that greatly increase the rate at which a chemical reaction occurs. The sum total of the chemical reactions in a cell represents that cell's metabolism.

细胞就像微型化工厂一样行使功能。即使最简单的细菌细胞也能进行成百上千种不同的化学转化，而这些化学变化在非生命界并不能如此高速地发生。几乎所有发生在细胞内的化学变化都需要酶的参与以大幅度地提高化学反应发生的速度。细胞内所有化学反应的总和就是细胞的新陈代谢。

Professional Words and Phrases

[1] **isolate** ['aisəleit] vt. 使分离; n. 分离物
[2] **culture** ['kʌltʃə] v. 培养; n. 培养, 培养物
[3] **cell division** 细胞分裂
[4] **sample** ['sæmpl] n. 样品, 样本; vt. 采样, 取样
[5] **in vitro** 在试管内, 体外
[6] **subcellular** [sʌb'seljulə] adj. 亚细胞的
[7] **electron microscope** 电子显微镜
[8] **macromolecule** [mækrə'mɔləkjuːl] n. 大分子
[9] **genetic program** 遗传程序
[10] **chromosome** ['krəuməsəum] n. 染色体
[11] **cell nucleus** 细胞核
[12] **genetic material** 遗传物质
[13] **genetic information** 遗传信息
[14] **oocyte** ['əuəsait] n. 卵母细胞
[15] **electromagnetic radiation** 电磁辐射
[16] **pigment** ['pigmənt] n. 色素
[17] **sucrose** ['sjuːkrəus] n. 蔗糖
[18] **starch** [staːtʃ] n. 淀粉
[19] **glucose** ['gluːkəus] n. 葡萄糖
[20] **chemical energy** 化学能
[21] **chemical reaction** 化学反应
[22] **multicellular** [mʌlti'seljulə] adj. 多细胞的
[23] **receptor** [ri'septə] n. 受体
[24] **growth factor** 生长因子
[25] **metabolic activity** 代谢活动
[26] **regulatory mechanism** 调控机制
[27] **mutation** [mjuː'teiʃən] n. 突变
[28] **secretion** [si'kriːʃən] n. 分泌
[29] **contraction** [kən'trækʃən] n. 收缩, 痉挛
[30] **fiber** ['faibə] n. 纤维

Exercises

1. Matching

1) cell division   a) any of the rod-shaped or threadlike DNA-containing structures of

cellular organisms that are located in the nucleus of eukaryotes, are usually ring-shaped in prokaryotes (as bacteria), and contain all or most of the genes of the organism

2) subcellular b) the parts of a cell that carry information that can be inherited, e.g. DNA, genes or chromosomes

3) multicellular c) a type of microscope that uses a particle beam of electrons to illuminate the specimen and produce a magnified image

4) genetic program d) of or related to more than two cells

5) chromosome e) a naturally occurring substance capable of stimulating cellular growth, proliferation and cellular differentiation

6) genetic material f) the process of segregating, elaborating, and releasing some material either functionally specialized or isolated for excretion

7) electron microscope g) the process by which a parent cell divides into two or more daughter cells

8) pigment h) of less than cellular scope or level of organization

9) oocyte i) a physiological change brought about by a temporal pattern of activation of a particular subset of genes

10) receptor j) a relatively permanent change in hereditary material involving either a physical change in chromosome relations or a biochemical change in the codons that make up genes

11) growth factor k) a coloring matter in animals and plants especially in a cell or tissue

12) mutation l) a chemical group or molecule (as a protein) on the cell surface or in the cell interior that has an affinity for a specific chemical group, molecule, or virus

13) secretion m) an egg before maturation

2. True or False
1) Cells can be removed from a plant or animal and cultured in a laboratory.
2) A single-celled protist can move away from an object in its path or moves toward a source of nutrients.
3) The bacterial cells can not carry out hundreds of different chemical transformations.
4) A breakdown in a cell's growth control can transform the cell into a cancer cell.
5) In any case, the two daughter cells produced by division have approximately equal volume.

3. Reading Comprehension
1) Which statement is not related to the cell complexity and ordered organization?
   A. The organization of atoms into small-sized molecules.
   B. The organization of the molecules into giant polymers.
   C. Organisms are built according to information encoded in a collection of genes.
   D. The organization of different types of polymeric molecules into complexes, which in turn are organized into subcellular organelles and finally into cells.
2) Which is responded by the receptors of the cells?
   A. hormone

    B. growth factor
    C. extracellular materials
    D. All
  3) Which of the following statements is not true?
    A. Cells may respond to specific stimuli by altering their metabolic activities.
    B. The first culture of human cells was begun by George Gey of Johns Hopkins University in 1951.
    C. Cells within a multicellular plant or animal don't respond to stimuli.
    D. Light energy is converted by photosynthesis into chemical energy.

4. Translation from English to Chinese

We are gradually learning more and more about how the cell controls its activities, but much more is left to discover. In the cell, the information for product design resides in the nucleic acids, and the construction workers are primarily proteins. It is the presence of these two types of macromolecules that, more than any other factor, sets the chemistry of the cell apart from that of the nonliving world. In the cell, the workers must act without the benefit of conscious direction. Each step of a process must occur spontaneously in such a way that the next step is automatically triggered. All the information required to direct a particular activity—whether it be the synthesis of a protein, the secretion of a hormone, or the contraction of a muscle fiber—must be present within the system itself.

5. Translation from Chinese to English

一些细胞对刺激的响应非常明显，例如，单细胞原生动物能远离其路径上的物体或径直向营养源移动。多细胞植物或动物的细胞对刺激的响应不十分明显，但它们也是有响应的。多数细胞是由与环境高度特异性地相互作用的受体所包被。细胞具有能响应激素、生长因子、细胞外物质以及其他细胞表面物质的受体。细胞受体能为激发目的细胞特定反应的外源物质提供通道供其通过。细胞可以通过改变代谢活动、为细胞分裂做准备、从一个地方移到另一个地方甚至采用自杀的方式响应特定的刺激。

3.3　The Concepts in Mammalian Cell Culture—Additional Reading

Cell culture[1] is the complex process by which cells are grown under controlled conditions. In practice, the term "cell culture" has come to refer to the culturing of cells derived from multicellular eukaryotes, especially animal cells. The concepts in mammalian cell culture are shown as below:

1) Isolation of cells

Cells can be isolated from tissues for ex vivo[2] culture in several ways. Cells can be easily purified from blood, however only the white cells are capable of growth in culture. Mononuclear[3] cells can be released from soft tissues[4] by enzymatic digestion[5] with enzymes such as collagenase[6], trypsin[7], or pronase[8], which break down the extracellular matrix[9]. Alternatively, pieces of tissue can be placed in growth media[10], and the cells that grow out are available for culture. This method is known as explant culture.

Cells that are cultured directly from a subject are known as primary cells. With the exception of some derived from tumors, most primary cell cultures have limited lifespan. After a certain number of population doublings (called the Hayflick limit[11]) cells undergo the process of senescence[12] and stop dividing, while generally retaining viability[13].

2) Maintaining cells in culture

Cells are grown and maintained at an appropriate temperature and gas mixture (typically, 37°C, 5% CO_2 for mammalian cells) in a cell incubator[14]. Culture conditions vary widely for each cell type, and variation of conditions for a particular cell type can result in different phenotypes[15] being expressed.

Aside from temperature and gas mixture, the most commonly varied factor in culture systems is the growth medium. Recipes for growth media can vary in pH, glucose concentration, growth factors, and the presence of other nutrients. The growth factors used to supplement media are often derived from animal blood, such as calf serum[16]. One complication of these blood-derived ingredients is the potential for contamination of the culture with viruses or prions, particularly in biotechnology medical applications.

Plating density[17] (number of cells per volume of culture medium) plays a critical role for some cell types. For example, a lower plating density makes granulosa[18] cells exhibit estrogen[19] production, while a higher plating density makes them appear as progesterone[20]—producing theca[21] lutein[22] cells.

Cells can be grown in suspension or adherent cultures. Some cells naturally live in suspension, without being attached to a surface, such as cells that exist in the bloodstream. There are also cell lines that have been modified to be able to survive in suspension cultures so that they can be grown to a higher density than adherent conditions would allow. Adherent cells require a surface, such as tissue culture plastic or microcarrier[23], which may be coated with extracellular matrix components to increase adhesion properties and provide other signals needed for growth and differentiation[24]. Most cells derived from solid tissues are adherent. Another type of adherent culture is organotypic[25] culture[26] which involves growing cells in a three-dimensional environment as opposed to two-dimensional culture dishes. This 3D culture system is biochemically[27] and physiologically[28] more similar to in vivo[29] tissue, but is technically challenging to maintain because of many factors (e.g. diffusion).

3) Cell line cross-contamination

Cell line[30] cross-contamination[31] can be a problem for scientists working with cultured cells. Studies suggest that anywhere from 15%～20% of the time, cells used in experiments have been misidentified or contaminated with another cell line. Problems with cell line cross contamination have even been detected in lines from the NCI-60 panel, which are used routinely for drug-screening[32] studies. Major cell line repositories[33] including the American Type Culture Collection (ATCC) and the German Collection of Microorganisms and Cell Cultures (DSMZ) have received cell line submissions[34] from researchers that were misidentified by the researcher. Such contamination poses a problem for the quality of research produced using cell culture lines, and the major repositories are now authenticating all cell line submissions. ATCC uses short tandem repeat (STR)[35] DNA fingerprinting[36] to authenticate its cell lines.

To address this problem of cell line cross-contamination, researchers are encouraged to authenticate their cell lines at an early passage[37] to establish the identity of the cell line. Authentication should be repeated before freezing cell line stocks, every two months during active culturing and before any publication of research data generated using the cell lines. There are many methods for identifying cell lines including isoenzyme[38] analysis[39], human lymphocyte[40] antigen[41] (HLA) typing[42], chromosomal[43] analysis[44], karyotyping[45], Morphology and STR analysis. One significant cell-line cross contaminant is the immortal HeLa cell line.

4) Manipulation of cultured cells

As cells generally continue to divide[46] in culture, they generally grow to fill the available area or volume. This can generate several issues: ①Nutrient depletion in the growth media. ②Accumulation of apoptotic[47]/necrotic[48] (dead) cells. ③Cell-to-cell contact can stimulate cell cycle[49] arrest[50], causing cells to stop dividing known as contact inhibition[51] or senescence. ④Cell-to-cell contact can stimulate cellular differentiation[52].

Among the common manipulations carried out on culture cells are media changes, passaging cells, and transfecting[53] cells. These are generally performed using tissue culture[54] methods that rely on sterile technique[55]. Sterile technique aims to avoid contamination with bacteria, yeast[56], or other cell lines. Manipulations are typically carried out in a biosafety cabinet[57] or laminar flow hood[58] to exclude contaminating micro-organisms. Antibiotics (e.g. penicillin[59] and streptomycin[60]) and antifungals[61] (e.g. amphotericin[62] B) can also be added to the growth media. As cells undergo metabolic processes[63], acid[64] is produced and the pH decreases. Often, a pH indicator[65] is added to the medium in order to measure nutrient depletion.

5) Passaging cells

Passaging (also known as subculture[66] or splitting cells[67]) involves transferring a small number of cells into a new vessel. Cells can be cultured for a longer time if they are split regularly, as it avoids the senescence associated with prolonged high cell density. Suspension cultures are easily passaged with a small amount of culture containing a few cells diluted[68] in a larger volume of fresh media. For adherent cultures, cells first need to be detached; this is commonly done with a mixture of trypsin-EDTA, however other enzyme mixes are now available for this purpose. A small number of detached cells can then be used to seed a new culture.

6) Transfection and transduction

Another common method for manipulating cells involves the introduction of foreign DNA by transfection. This is often performed to cause cells to express a protein of interest. More recently, the transfection of RNAi constructs[69] has been realized as a convenient mechanism for suppressing the expression of a particular gene/protein. DNA can also be inserted into cells using viruses, in methods referred to as transduction[70], infection[71] or transformation[72]. Viruses, as parasitic[73] agents, are well suited to introduce DNA into cells, as this is a part of their normal course of reproduction.

Notes to the Difficult Sentences

To address this problem of cell line cross-contamination, researchers are encouraged to authenticate their cell lines at an early passage to establish the identity of the cell line.

Authentication should be repeated before freezing cell line stocks, every two months during active culturing and before any publication of research data generated using the cell lines. There are many methods for identifying cell lines including isoenzyme analysis, human lymphocyte antigen (HLA) typing, Chromosomal analysis, Karyotyping, Morphology and STR analysis.

为了解决细胞交叉污染的问题，鼓励研究人员在传代早期验证他们的细胞系以证实细胞系的身份。该验证在冻存细胞系之前、在活化期间的每两个月以及在发表使用该细胞系所获得的研究数据之前需要进行重复。有很多鉴定细胞系的方法，其中包括同工酶分析、人淋巴细胞（HLA）分型、染色体分析、核型分析以及形态学和短串联重复（STR）分析。

Among the common manipulations carried out on culture cells are media changes, passaging cells, and transfecting cells. These are generally performed using tissue culture methods that rely on sterile technique. Sterile technique aims to avoid contamination with bacteria, yeast, or other cell lines. Manipulations are typically carried out in a biosafety hood or laminar flow cabinet to exclude contaminating micro-organisms. Antibiotics (e.g. penicillin and streptomycin) and antifungals (e.g. Amphotericin B) can also be added to the growth media.

培养细胞的基本操作有更换培养基、细胞传代和细胞转染。这些操作都是在无菌操作条件下使用组织培养方法进行的。无菌操作技术目的是避免细菌、酵母或其他细胞系的污染。该操作正常是在生物安全柜或层流净化罩内进行的，为的是排除微生物的污染。当然也可以在生长培养基中加入抗生素（如青霉素和链霉素）和抗真菌剂（如两性霉素B）。

Professional Words and Phrases

[1] **cell culture** 细胞培养
[2] **ex vivo** （与培养有关的）体外
[3] **mononuclear** ['mɔnəu 'njuːkliə] adj. 单核的
[4] **soft tissue** 软组织
[5] **enzymatic digestion** 酶消化
[6] **collagenase** ['kɔlədʒəneis] n. 胶原酶
[7] **trypsin** ['tripsin] n. 胰蛋白酶
[8] **pronase** ['prəuneis] n. 链霉蛋白酶
[9] **matrix** ['meitriks] n. 基质
[10] **media** ['miːdiə] n. 培养基，medium 的复数形式
[11] **Hayflick limit** 海弗利克极限（指培养中细胞生命的自然极限）
[12] **senescence** [sə'nesəns] n. 衰老
[13] **viability** [vaiə'biliti] n. 生存能力，发育能力
[14] **incubator** ['inkjubeitə] n. 培养箱
[15] **phenotype** ['fiːnətaip] n. 表型
[16] **serum** ['siərəm] n. 血清
[17] **plating density** 接种密度
[18] **granulosa** [grænju'ləusə] n. 粒层细胞
[19] **estrogen** ['estrədʒən] n. 雌激素
[20] **progesterone** [prəu'dʒestərəun] n. 黄体酮
[21] **theca** ['θiːkə] n. 泡膜

[22] **lutein** [ˈluːtiin] n. 黄体素, 叶黄素
[23] **microcarrier** [maikəuˈkæriə] n. 微载体
[24] **differentiation** [difərenʃiˈeiʃən] n. 分化
[25] **organotypic** [ˈɔːgənəuˈtipik] adj. 器官型的
[26] **organotypic culture** 器官型培养
[27] **biochemically** [ˈbaiəˈkemikəli] adv. 在生物化学上
[28] **physiologically** [fiziəˈlɔdʒikəli] adv. 在生理学上
[29] **in vivo** 体内
[30] **cell line** 细胞系
[31] **cross-contamination** 交叉污染
[32] **drug-screening** 药物筛选
[33] **repository** [riˈpɔzətəuri] n. 容器, 仓库, 贮藏室
[34] **submission** [sʌbˈmiʃən] n. 投（稿）, 提交
[35] **short tandem repeat (STR)** 短串联重复
[36] **DNA fingerprinting** DNA 指纹图谱
[37] **passage** [ˈpæsidʒ] n. & v. 传代
[38] **isoenzyme** [aisəuˈenzaim] n. 同工酶
[39] **isoenzyme analysis** 同工酶分析
[40] **lymphocyte** [ˈlimfəsait] n. 淋巴细胞
[41] **antigen** [ˈæntədʒən] n. 抗原
[42] **human lymphocyte antigen typing** 人淋巴细胞抗原分型
[43] **chromosomal** [ˈkrəuməsəuməl] adj. 染色体的
[44] **chromosomal analysis** 染色体分析
[45] **karyotyping** [ˈkæriətaipiŋ] n. 染色体组型分型
[46] **divide** [diˈvaid] vi. 分裂, 分开
[47] **apoptotic** [æpəˈtəutik] adj. 细胞凋亡的
[48] **necrotic** [neˈkrɔtik] adj. 坏死的, 坏疽的
[49] **cell cycle** 细胞周期
[50] **cell cycle arrest** 细胞周期阻滞
[51] **contact inhibition** 接触抑制
[52] **cellular differentiation** 细胞分化
[53] **transfect** [trænsˈfekt] vt. 转染
[54] **tissue culture** 组织培养
[55] **sterile technique** 无菌操作技术
[56] **yeast** [jiːst] n. 酵母, 发酵剂
[57] **biosafety cabinet** 生物安全柜
[58] **laminar flow hood** 层流净化罩
[59] **penicillin** [peniˈsilin] n. 青霉素
[60] **streptomycin** [ˈstreptəˈmaisin] n. 链霉素
[61] **antifungal** [æntiˈfʌŋgəl] adj. 抗真菌的, 杀真菌的; n. 杀真菌剂
[62] **amphotericin** [æmfəˈterəsin] n. 两性霉素

[63] **metabolic process** 代谢过程
[64] **acid** [ˈæsid] n. 酸, 酸性物质; adj. 酸的, 酸性的
[65] **pH indicator** pH 指示剂
[66] **subculture** [ˈsʌbkʌltʃə] n. 继代培养
[67] **splitting cells** 分细胞
[68] **dilute** [daiˈluːt] vt. 稀释, 冲淡
[69] **construct** [ˈkɔnstrʌkt] n. 构建物
[70] **transduction** [trænsˈdʌkʃən] n. 转导
[71] **infection** [inˈfekʃən] n. 传染, 侵染
[72] **transformation** [trænsfəˈmeiʃən] n. 转化
[73] **parasitic** [pærəˈsitik] adj. 寄生的

Exercises

1. Matching

1) cell culture — a) a permanently established cell culture that will proliferate indefinitely given appropriate fresh medium and space

2) collagenase — b) the process by which DNA is transferred from one bacterium to another by a virus

3) Hayflick limit — c) the genetic alteration of a cell resulting from the direct uptake, incorporation and expression of exogenous genetic material (exogenous DNA) from its surrounding and taken up through the cell membrane(s)

4) cell line — d) the culturing of cells derived from multicellular eukaryotes, especially animal cells

5) phenotype — e) a test to examine chromosomes in a sample of cells, which can help identify genetic problems as the cause of a disorder or disease

6) karyotyping — f) any observable characteristic or trait of an organism

7) cell cycle — g) enzymes that break the peptide bonds in collagen

8) sterile technique — h) the number of times a normal cell population will divide before it stops, presumably because the telomeres reach a critical length

9) transduction — i) the series of events that takes place in a cell leading to its division and duplication (replication)

10) transformation — j) also named aseptic technique, a procedure that is performed under sterile conditions

2. True or False

1) A higher plating density makes granulosa cells appear as progesterone-producing theca lutein cells.

2) Mammalian Cells are typically grown and maintained at 37°C, 5% CO_2 in a cell incubator.

3) More recently, the transfection of RNAi constructs has been realized as a convenient mechanism for suppressing the expression of a particular gene/protein.

4) DSMZ uses short tandem repeat (STR) DNA fingerprinting to authenticate its cell lines.

5) Cell-to-cell contact can't stimulate cellular differentiation.

3. Reading Comprehension

1) Which is not used to avoid the cell line contamination with micro-organisms?
   A. freezing cell line
   B. manipulations in a biosafety hood
   C. adding antibiotics
   D. adding antifungals

2) Which is not the method for identification of cell lines?
   A. isoenzyme analysis
   B. chromosomal analysis
   C. sterile technique
   D. morphology and STR analysis

3) Which statement is not true?
   A. A pH indicator is added to the medium in order to measure nutrient depletion.
   B. Some cells naturally live in suspension, without being attached to a surface.
   C. Adhere cultures are easily passaged with a small amount of culture containing a few cells diluted in a larger volume of fresh media.
   D. One significant cell-line cross contaminant is the immortal HeLa cell line.

4. Translation from English to Chinese

Cells can be grown in suspension or adherent cultures. Some cells naturally live in suspension, without being attached to a surface, such as cells that exist in the bloodstream. There are also cell lines that have been modified to be able to survive in suspension cultures so that they can be grown to a higher density than adherent conditions would allow. Adherent cells require a surface, such as tissue culture plastic or microcarrier, which may be coated with extracellular matrix components to increase adhesion properties and provide other signals needed for growth and differentiation. Most cells derived from solid tissues are adherent. Another type of adherent culture is organotypic culture which involves growing cells in a three-dimensional environment as opposed to two-dimensional culture dishes.

5. Translation from Chinese to English

另一种常规的细胞操作方法与通过转染引入外源基因有关。这通常是为了使细胞表达令人感兴趣的蛋白质。最近，转染 RNAi 的构建物已经被认为是抑制特定基因/蛋白表达的一种非常便捷的机制。DNA 也可通过病毒被插入细胞中，其方法被称作转染、侵染或转化。作为寄生病原体，病毒之所以非常适用于将 DNA 引入细胞，是因为这是病毒正常繁殖过程的一部分。

Chapter 4　Botany—Plant Biology

[本章中文导读]

　　植物学作为生物学的重要分支，是研究植物生命的科学。本章主要介绍了植物学的范围和重要性（第 4.1 节），植物的花、果实和种子（第 4.2 节），光合作用（第 4.3 节）以及转基因植物方面的一些成就（第 4.4 节）等与植物密切相关的知识点。作者想通过用英文讲解植物学的以上专业知识点，使学生能够初步掌握与植物学有关的专业英语词汇的准确含义和相关使用方法。

4.1　The Scope and Importance of Botany

　　Botany, plant science(s), or plant biology is a branch of biology that involves the scientific study of plant life. Botany covers a wide range of scientific disciplines concerned with the study of plants, algae and fungi, including structure, growth, reproduction, metabolism, development[1], diseases, chemical properties, and evolutionary[2] relationships among taxonomic groups. Botany began with early human efforts to identify edible, medicinal[3] and poisonous plants, making it one of the oldest sciences. Today botanists study over 550,000 species of living organisms. The term "botany" comes from Greek βοτάνη, meaning "pasture, grass, fodder", perhaps via the idea of a livestock keeper needing to know which plants are safe for livestock to eat.

　　As with other life forms in biology, plant life can be studied from different perspectives, from the molecular, genetic and biochemical[4] level through organelles, cells, tissues, organs, individuals[5], plant populations[6], and communities[7] of plants. At each of these levels a botanist might be concerned with the classification (taxonomy[8]), structure (anatomy[9] and morphology), or function (physiology) of plant life.

　　<u>Historically all living things were grouped as animals or plants, and botany covered all organisms not considered animals. Some organisms once included in the field of botany are no longer considered to belong to the plant kingdom—these include fungi (studied in mycology), lichens[10] (lichenology[11]), bacteria (bacteriology), viruses (virology) and single-celled algae, which are now grouped as part of the protista[12]. However, attention is still given to these groups by botanists, and fungi, lichens, bacteria and photosynthetic protists[13] are usually covered in introductory botany courses.</u>

　　The study of plants is vital because they are a fundamental part of life on Earth, which generates the oxygen, food, fibers, fuel and medicine that allow humans and other life forms to exist. Through photosynthesis, plants absorb carbon dioxide[14], a greenhouse[15] gas that in large amounts can affect global climate. Additionally, they prevent soil erosion[16] and are influential in the water cycle. A good understanding of plants is crucial to the future of human societies as it allows us to: ①Produce food to feed an expanding population. ②Understand fundamental life processes. ③Produce medicine and materials to treat diseases and other ailments. ④Understand environmental changes more clearly.

Notes to the Difficult Sentences

Historically all living things were grouped as animals or plants, and botany covered all organisms not considered animals. Some organisms once included in the field of botany are no longer considered to belong to the plant kingdom—these include fungi (studied in mycology), lichens (lichenology), bacteria (bacteriology), viruses (virology) and single-celled algae, which are now grouped as part of the protista. However, attention is still given to these groups by botanists, and fungi, lichens, bacteria and photosynthetic protists are usually covered in introductory botany courses.

在历史上，所有的生命体都被归类为动物或植物，而植物学所涉及的是被认为不是动物的所有生物。一些曾经被纳入植物学领域的生物现在已不再被认为属于植物界，它们是真菌学研究的真菌、地衣学研究的地衣、细菌学研究的细菌、病毒学研究的病毒以及现在被归入原生动物界的单细胞藻类。然而，植物学家仍然关注着这些生物类群，而且现在植物学导论的课程通常仍然包括真菌、地衣、细菌和能进行光合作用的原生动物等。

Professional Words and Phrases

[1]   **development**   [di'veləpmənt]   n. 发育
[2]   **evolutionary**   [evə'luːʃən.eri]   adj. 进化的
[3]   **medicinal**   [mə'disənəl]   adj. 药物的
[4]   **biochemical**   ['baiəu'kemikəl]   adj. 生物化学的; n. 生物化工
[5]   **individual**   [indi'vidjuəl]   n. 个体
[6]   **population**   [pɔpju'leiʃən]   n. 种群
[7]   **community**   [kə'mjuːniti]   n. 群落
[8]   **taxonomy**   [tæk'sɔnəmi]   n. 分类学
[9]   **anatomy**   [ə'nætəmi]   n. 解剖学，解剖
[10] **lichen**   ['laikin]   n. 地衣，青苔
[11] **lichenology**   [laikə'nɔlədʒi]   n. 地衣学
[12] **protista**   [prəu'tistə]   n. 原生动物界
[13] **protist**   ['prəutist]   n. 原生生物
[14] **carbon dioxide**   二氧化碳
[15] **greenhouse**   ['griːnhaus]   n. 温室
[16] **erosion**   [i'rəuʒən]   n. 腐蚀，侵蚀

Exercises

1. Matching

1) individual   a) a group of interacting organisms sharing a populated environment
2) population   b) a process that converts carbon dioxide into organic compounds, especially sugars, using the energy from sunlight

3) community   c) a single organism as distinguished from a group

4) taxonomy   d) all the organisms that both belong to the same species and live in the same geographical area

5) anatomy   e) a diverse group of eukaryotic microorganisms

6) protist   f) the practice and science of classification

7) photosynthesis   g) a branch of biology and medicine that is the consideration of the structure of living things

2. True or False

1) Plant life can be studied from different perspectives, from the molecular, genetic and biochemical level.

2) Botany began with early human efforts to identify edible, medicinal and poisonous plants.

3) Fungi, lichens, bacteria, viruses and single-celled algae were not included in the plant kingdom.

3. Reading Comprehension

1) Which is not considered by a botanist?
   A. the classification of plant life
   B. the structure of plant life
   C. the function of plant life
   D. the classification of algae

2) Which is not included in the Botany now?
   A. photosynthetic protists
   B. plant physiology
   C. plant taxonomy
   D. plant structure

3) Which of the following statements is not true?
   A. A good understanding of plants is crucial to the future of human societies.
   B. Botany is considered to be one of the oldest sciences.
   C. Botany courses don't cover the knowledge of fungi, lichens, bacteria now.
   D. Today botanists study over 550,000 species of living organisms.

4. Translation from English to Chinese

Botany, plant science(s), or plant biology is a branch of biology that involves the scientific study of plant life. Botany covers a wide range of scientific disciplines concerned with the study of plants, algae and fungi, including structure, growth, reproduction, metabolism, development, diseases, chemical properties, and evolutionary relationships among taxonomic groups. Botany began with early human efforts to identify edible, medicinal and poisonous plants, making it one of the oldest sciences.

5. Translation from Chinese to English

对植物的研究是至关重要的，这是因为它们是地球上一个重要的生命组成部分，它们产生人和其他生命生存所需要的氧、食物、纤维、能源和药物。植物通过光合作用吸收二氧化碳，这是一种大量存在并能影响全球气候的温室气体。此外，植物可以防止土壤被侵蚀并且影响水循环。深入地理解植物对人类社会的未来至关重要，这是因为它能使我们：①生产供养不断增加的人口的食物。②理解重要的生命过程。③生产治疗疾病和其他不适的药物以及材料。④更加清楚地理解环境变化。

4.2 Flowers, Fruits and Seeds of Plants

Flowers

Flowers arise from apical meristems[1] similar to vegetative shoots[2] but, unlike them, have determinate growth. The floral primordia[3] develop into four different kinds of specialized leaves that are borne in whorls[4] at the tip of the stem. The two outer whorls are sterile, while the inner two are fertile. The first formed outer whorl—the calyx[5]—is the most leaflike and its individual parts, the sepals[6], often are green. The petals[7] of the next whorl, the corolla[8], frequently are brightly colored and in a majority of flowers retain some semblance to leaves. (Together the calyx and the corolla are called the perianth[9].) The next two whorls, the androecium[10] and the gynoecium[11], are composed of highly modified reproductive structures that have lost their leaflike appearance. The androecium is composed of stamens[12] and the gynoecium of carpels[13]. (Pistil[14] is sometimes used as the term for a single carpel or a group of fused carpels.) The stamens are microsporophylls[15] and have a stalk[16], the filament[17], at the top of which the pollen[18]-bearing anthers[19] are located. A carpel is a megasporophyll[20] and has as its base an enlarged ovary[21] from which the style bearing a stigma[22] arises. The whorls are attached to the receptacle area at the end of the flower stalk or pedicel[23]. Some flowers arise singly, but more are produced and arranged in groups called inflorescences[24]. The stalk of an inflorescence is the peduncle[25] and the extension of the axis in the inflorescence is the rachis[26], to which the pedicels of the individual flowers are attached.

Fruits

Fruits are a uniquely angiosperm[27] feature: part of the pollinated[28] flower ripens and becomes the fruit. Two types of fruits with a generally different mechanism of seed distribution exist: dehiscent[29] and nondehiscent[30] fruits. Dehiscent fruits set the seeds free by opening of the fruit that remains itself with the mother plant. In contrast, nondehiscent fruits and their seeds are a dispersal unit. Dehiscent fruit types are follicles[31], pulses, pods[32] and capsules[33]. Nondehiscent fruits are berries[34], drupes[35] and nuts.

In the simplest case, the seeds are set free from the ovary after ripening by opening of the fruit (dehiscent fruits). But very often, both fruits and seeds form an integrated part of the dispersal unit (nondehiscent fruit). If this happens, the tissue of the ovary is strongly differentiated. The pericarp[36] is the part of a fruit that is formed by the ripening of the ovary wall. It is organized into three layers: the skin (exocarp[37]), the often fleshy middle (mesocarp[38]), and the membranous or stony inner layer (endocarp[39]). These layers may become skin-like and leathery, fleshy or stringy (sclerotic) during fructification. The fruits produce gadgets for the protection and the dispersal of the seeds. Additionally, other parts of the flower may contribute to the formation of the fruit, like the flower axis or the calyx. Depending on the organization of the gynoecium, single fruits or multiple fruits develop.

Seeds

Seeds are normally surrounded by a tough shell (testa[40]) that develops from the integument[41] (or several integuments) and that protects a largely developed embryo and its nutriments. The embryo is thus preserved from loss of water and bad weather. The nutriments[42] are in a special nourishing tissue, the endosperm[43] or in the perisperm[44] (developed from tissue of the nucellus[45] as with pepper). They have either to be made useable and to be used during germination[46] or they are already stored in the cotyledons[47] (beans, peas, almonds). Monocotyledons[48] have normally only one cotyledon, dicotyledons[49] have two and gymnosperms[50] several.

<u>The number of seeds that is produced by a single plant in the course of its life varies enormously. Some species like the coconut palm produce only few but very well provided descendants while others produce up to several million seeds that are inevitably provided rather badly.</u> Numerous devices support the dispersal of the descendants since only a suitable habitat enables the seed to develop into a viable plant. If the dispersal is achieved only through devices that the plant itself produces, it is spoken of self-dispersal or autochory. If extern factors such as wind, water, animals, etc. are involved, then the mode of dispersal is called allochor. Seeds are dispersed either by self-dispersal, also called autochory[51], or by allochory[52] which means that extern factors are involved. Allochor modes of dispersal are dispersal by wind (anemochorys[53]), water (hydrochory[54]) or by animals (zoochory[55]).

Notes to the Difficult Sentences

The first formed outer whorl—the calyx—is the most leaflike and its individual parts, the sepals, often are green. The petals of the next whorl, the corolla, frequently are brightly colored and in a majority of flowers retain some semblance to leaves. (Together the calyx and the corolla are called the perianth.) The next two whorls, the androecium and the gynoecium, are composed of highly modified reproductive structures that have lost their leaf-like appearance. The androecium is composed of stamens and the gynoecium of carpels.

最先产生的外部轮生体是花萼，它最像叶子，而且其单独个体萼片经常是绿色的。第二个轮生体是花冠，它的花瓣颜色通常是鲜亮的，且是花的主要部分，保留了一些与叶子的相似之处。（花萼和花冠统称为花被。）后两个轮生体，即雄蕊群和雌蕊群，是由高度修饰的生殖结构组成的，这些结构已经失去叶子的特征。雄蕊群由雄蕊组成，而雌蕊群则由心皮组成。

The number of seeds that is produced by a single plant in the course of its life varies enormously. Some species like the coconut palm produce only few but very well provided descendants while others produce up to several million seeds that are inevitably provided rather badly.

单一的植株在生命过程中产生的种子数量差异是很显著的。像椰子棕榈这样的物种产生非常少的种子，但却能保证其传代无忧，而其他物种产生数量高达几百万的种子，这些种子中的一些可能就很差。

Professional Words and Phrases

[1] meristem ['meristem] n. 分生组织
[2] shoots [ʃuːts] n. 嫩枝
[3] primordia [praiˈmɔːdjə] n. 原基，原始细胞, primordium 的复数形式
[4] whorl [wəːl] n. 轮生体
[5] calyx ['keiliks] n. 花萼
[6] sepal ['siːpəl] n. 萼片
[7] petal ['petl] n. 花瓣
[8] corolla [kəˈrɔlə] n. 花冠
[9] perianth ['periænθ] n. 花被
[10] androecium [ænˈdriʃiəm] n. 雄蕊群
[11] gynoecium [dʒiˈniːsiəm] n. 雌蕊群
[12] stamen ['steimən] n. 雄蕊
[13] carpel ['kɑːpəl] n. 心皮，雌蕊叶
[14] pistil ['pistil] n. 雌蕊
[15] microsporophyll [maikrəˈspɔːrəfil] n. 小孢子叶
[16] stalk [stɔːk] n. 茎，梗
[17] filament ['filəmənt] n. 花丝
[18] pollen ['pɔlən] n. 花粉
[19] anther ['ænθə] n. 花粉囊，花药
[20] megasporophyll [megəˈspɔːrəfil] n. 大孢子叶
[21] ovary ['əuvəri] n. 卵巢，子房
[22] stigma ['stigmə] n. 柱头
[23] pedicel ['pedəsəl] n. 花梗，肉茎
[24] inflorescence [infləuˈresəns] n. 花序
[25] peduncle [piˈdʌŋkl] n. 花梗，肉茎
[26] rachis ['reikis] n. 花轴
[27] angiosperm [ˌændʒiəˌspəːm] n. 被子植物
[28] pollinate ['pɔlineit] vt. 给……授粉
[29] dehiscent [diˈhisənt] adj. 裂开性的
[30] nondehiscent [nʌndiˈhisənt] adj. 不裂开性的
[31] follicle ['fɔlikl] n. 骨突果，卵泡细胞
[32] pod [pɔd] n. 荚果
[33] capsule ['kæpsjuːl] n. 蒴果，荚膜
[34] berry ['beri] n. 浆果
[35] drupe [druːp] n. 核果
[36] pericarp ['perikɑːp] n. 果皮
[37] exocarp ['eksəukɑːp] n. 外果皮
[38] mesocarp ['mesəkɑːp] n. 中果皮

[39] **endocarp** ['endəukɑ:p] n. 内果皮
[40] **testa** ['testə] n. 外种皮
[41] **integument** [in'tegjumənt] n. 外皮
[42] **nutriment** ['nju:trimənt] n. 营养品，养料，滋养物
[43] **endosperm** ['endəspə:m] n. 胚乳
[44] **perisperm** ['perispə:m] n. 外胚乳
[45] **nucellus** [nu:'seləs] n. 珠心
[46] **germination** [dʒə:mi'neiʃən] n. 发芽，萌芽
[47] **cotyledon** [kɔtəl'i:dən] n. 子叶
[48] **monocotyledon** [mɔnəkɔtəl'i:dən] n. 单子叶植物
[49] **dicotyledon** [daikɔtəl'i:dən] n. 双子叶植物
[50] **gymnosperm** ['dʒimnəspə:m] n. 裸子植物
[51] **autochory** [ɔ:təu'tʃɔri] n. 自动散播
[52] **allochory** [ælə'tʃɔri] n. 异地传播
[53] **anemochory** [ə'ni:məkɔ:ri] n. 风力传播
[54] **hydrochory** ['haidrəkɔ:ri] n. 水流传播
[55] **zoochory** ['zəuəkəri] n. 动物传播

Exercises

1. Matching

1) meristem   a) typically the innermost whorl of structures in a flower
2) androecium   b) a significant part of the embryo within the seed of a plant
3) gynoecium   c) the tissue in most plants consisting of undifferentiated cells
4) microsporophyll   d) the collection of stamens in a flower
5) megasporophyll   e) the carpel of a flowering plant
6) endosperm   f) the stamen of a flowering plant
7) cotyledon   g) the tissue produced inside the seeds of most flowering plants around the time of fertilization

2. True or False

1) The floral primordia develop into four different kinds of specialized leaves that are borne in whorls at the tip of the stem.
2) The sepals of the next whorl, the corolla, frequently are brightly colored and in a majority of flowers retain some semblance to leaves.
3) The androecium is composed of carpels and the gynoecium of stamens.
4) Berries, drupes and nuts set the seeds free by opening of the fruit that remains itself with the mother plant.
5) Dicotyledons have normally only one cotyledon, while monocotyledons have two.

3. Reading Comprehension

1) Which of the following whorls is sterile?
   A. calyx
   B. the androecium

C. the gynoecium

   D. none of the above

2) Which is not included in the dehiscent fruit types?

   A. follicles

   B. pulses

   C. berries

   D. pods

3) Which does not belong to allochor modes of dispersal?

   A. anemochory

   B. hydrochory

   C. zoochory

   D. autochory

4. Translation from English to Chinese

   The pericarp is the part of a fruit that is formed by the ripening of the ovary wall. It is organized into three layers: the skin (exocarp), the often fleshy middle (mesocarp), and the membranous or stony inner layer (endocarp). These layers may become skin-like and leathery, fleshy or stringy (sclerotic) during fructification. The fruits produce gadgets for the protection and the dispersal of the seeds. Additionally, other parts of the flower may contribute to the formation of the fruit, like the flower axis or the calyx.

5. Translation from Chinese to English

   只有合适的环境才能使种子发育成有生命力的植物，因此植物就不得不借助各种传播方法帮助其将后代传播到合适的环境中去。如果传播是通过植物自身产生的方法实现的，该方法称为自动传播。如果有风、水、动物等其他外界因子参与，那么这种传播方法就称为媒介传播。种子可以通过采用自动传播的方式传播，也可以通过外界因子采用媒介传播的方式传播。媒介传播指的是借助风（风力传播）、水（水流传播）或是动物（动物传播）等进行传播的方式。

4.3　Photosynthesis

   Photosynthesis is the process of converting light energy to chemical energy and storing it in the bonds[1] of sugar. This process occurs in plants and some algae (Kingdom Protista). Plants need only light energy, CO_2, and H_2O to make sugar. The process of photosynthesis takes place in the chloroplasts[2], specifically using chlorophyll[3], the green pigment involved in photosynthesis.

   Photosynthesis takes place primarily in plant leaves, and little to none occurs in stems[4], etc. The parts of a typical leaf include the upper and lower epidermis[5], the mesophyll[6], the vascular bundle(s)[7] (veins[8]), and the stomates[9]. The upper and lower epidermal cells[10] do not have chloroplasts, thus photosynthesis does not occur there. They serve primarily as protection for the rest of the leaf. The stomates are holes which occur primarily in the lower epidermis and are for air exchange: they let CO_2 in and O_2 out. The vascular bundles or veins in a leaf are part of the plant's transportation system, moving water and nutrients around the plant as needed. The mesophyll cells[11] have chloroplasts and this is where photosynthesis occurs.

As you hopefully recall, the parts of a chloroplast include the outer and inner membranes, intermembrane space[12], stroma[13], and thylakoids[14] stacked in grana[15]. The chlorophyll is built into the membranes of the thylakoids.

Chlorophyll looks green because it absorbs red and blue light, making these colors unavailable to be seen by our eyes. It is the green light which is not absorbed that finally reaches our eyes, making chlorophyll appear green. However, it is the energy from the red and blue light that are absorbed that is, thereby, able to be used to do photosynthesis. The green light we can see is not/cannot be absorbed by the plant, and thus cannot be used to do photosynthesis.

There are two parts to photosynthesis:

The light reaction[16] happens in the thylakoid membrane and converts light energy to chemical energy. This chemical reaction must, therefore, take place in the light. Chlorophyll and several other pigments such as beta-carotene[17] are organized in clusters in the thylakoid membrane and are involved in the light reaction. Each of these differently-colored pigments can absorb a slightly different color of light and pass its energy to the central chlorphyll molecule to do photosynthesis. The central part of the chemical structure[18] of a chlorophyll molecule is a porphyrin[19] ring, which consists of several fused rings of carbon and nitrogen with a magnesium[20] ion in the center.

The dark reaction[21] takes place in the stroma within the chloroplast, and converts CO_2 to sugar. This reaction doesn't directly need light in order to occur, but it does need the products of the light reaction (ATP and another chemical called NADPH). The dark reaction involves a cycle called the Calvin cycle[22] in which CO_2 and energy from ATP are used to form sugar. Actually, notice that the first product of photosynthesis is a three-carbon compound[23] called glyceraldehyde 3-phosphate[24]. Almost immediately, two of these join to form a glucose molecule.

Notes to the Difficult Sentences

The parts of a typical leaf include the upper and lower epidermis, the mesophyll, the vascular bundle(s) (veins), and the stomates. The upper and lower epidermal cells do not have chloroplasts, thus photosynthesis does not occur there. They serve primarily as protection for the rest of the leaf. The stomates are holes which occur primarily in the lower epidermis and are for air exchange: they let CO_2 in and O_2 out. The vascular bundles or veins in a leaf are part of the plant's transportation system, moving water and nutrients around the plant as needed. The mesophyll cells have chloroplasts and this is where photosynthesis occurs.

叶子典型的组成部分包括上表皮和下表皮、叶肉、维管束（叶脉）和气孔。上下表皮细胞不含叶绿体，因此这里不能进行光合作用。它们的功能主要是保护叶子的其余部分。气孔主要出现在下表皮，是用于气体交换的：它们让二氧化碳进入并排出氧。叶片上的维管束或叶脉是植物转运系统的组成部分，它负责将植物所必需的水和营养物质运输到植物的各部分。叶肉细胞含有叶绿体，这就是光合作用的场所。

Professional Words and Phrases

[1] **bond** [bɔnd] n. 共价键

[2] **chloroplast** ['klɔ:rəplæst] n. 叶绿体
[3] **chlorophyll** ['klɔrəfil] n. 叶绿素
[4] **stem** [stem] n. 茎
[5] **epidermis** [,epi'də:mis] n. 表皮, 上皮
[6] **mesophyll** ['mesəfil] n. 叶肉
[7] **vascular bundle** 维管束
[8] **vein** [vein] n. 静脉, 叶脉
[9] **stomate** ['stəumeit] n. 气孔
[10] **epidermal cell** 表皮细胞
[11] **mesophyll cell** 叶肉细胞
[12] **intermembrane space** 膜间腔
[13] **stroma** ['strəumə] n. 基质, 子座
[14] **thylakoid** ['θailəkɔid] n. 类囊体
[15] **grana** ['greinə] n. 基粒
[16] **light reaction** 光反应
[17] **beta-carotene** [beitə'kærətin] n. β-胡萝卜素
[18] **chemical structure** 化学结构
[19] **porphyrin** ['pɔ:fərin] n. 卟啉
[20] **magnesium** ['mæg'ni:ziəm] n. 镁
[21] **dark reaction** 暗反应
[22] **Calvin cycle** 卡尔文循环
[23] **three-carbon compound** 三碳化合物
[24] **glyceraldehyde 3-phosphate** 3-磷酸甘油醛

Exercises

1. Matching

1) chloroplast   a) a part of the transport system in vascular plants
2) epidermis   b) the fluid in between grana, where carbohydrate formation reactions occur in the chloroplasts of plant cells photosynthesizing
3) vascular bundle c) a strongly-coloured red-orange pigment abundant in plants and fruits
4) stomate   d) the green pigment involved in photosynthesis
5) stroma   e) holes which occur primarily in the lower epidermis and are for air exchange
6) thylakoid   f) an organelle found in plant cells and other eukaryotic organisms that conduct photosynthesis
7) chlorophyll   g) the outer layer of the skin
8) beta-carotene   h) a membrane-bound compartment inside chloroplasts and cyanobacteria

2. True or False

1) Photosynthesis is the process of converting light energy to chemical energy, only taking place in plants.
2) Photosynthesis takes place primarily in plant upper and lower epidermal cells.

3) Chlorophyll looks green because it absorbs red and blue light, not absorb the green light.

4) The light reaction happens in the thylakoid membrane and converts light energy to chemical energy.

5) The dark reaction takes place in the stroma within the chloroplast, and converts CO_2 to sugar.

3. Reading Comprehension

1) Which is not a substrate for the photosythesis?

   A. light energy

   B. CO_2

   C. O_2

   D. H_2O

2) Which of the following statements is not true?

   A. Photosynthesis only takes place primarily in plant leaves.

   B. The green light cannot be used to do photosynthesis.

   C. Chlorophyll and several other pigments such as beta-carotene are organized in clusters in the thylakoid membrane.

   D. ATP and NADPH are products of photosysthesis.

3) Where does the dark reaction happen?

   A. the outer and inner membranes

   B. intermembrane space

   C. stroma

   D. thylakoids

4. Translation from English to Chinese

The light reaction happens in the thylakoid membrane and converts light energy to chemical energy. This chemical reaction must, therefore, take place in the light. Chlorophyll and several other pigments such as beta-carotene are organized in clusters in the thylakoid membrane and are involved in the light reaction. Each of these differently-colored pigments can absorb a slightly different color of light and pass its energy to the central chlorphyll molecule to do photosynthesis.

5. Translation from Chinese to English

暗反应发生在叶绿体基质中，负责的是将二氧化碳转化成糖。该反应不需要光直接参与就能发生，但需要光反应的产物（ATP 和另一种叫 NADPH 的化学物质）。暗反应所涉及的一个循环称为卡尔文循环，在这个循环中二氧化碳和 ATP 的能量被用于糖的合成。

4.4 Some Achievements of Transgenic Plants—Additional Reading

Progress is being made on several fronts to introduce new traits[1] into plants using recombinant DNA technology. The genetic manipulation[2] of plants has been going on since the dawn of agriculture, but until recently this has required the slow and tedious process of crossbreeding[3] varieties. Genetic engineering promises to speed the process and broaden the scope of what can be done.

1) Improved Nutritional Quality

Milled rice is the staple food for a large fraction of the world's human population. Milling rice removes the husk and any beta-carotene it contained. Beta-carotene is a precursor to vitamin A, so it is not surprising that vitamin A deficiency is widespread, especially in the countries of Southeast Asia.

The synthesis of beta-carotene requires a number of enzyme-catalyzed[4] steps. In January 2000, a group of European researchers reported that they had succeeded in incorporating three transgenes[5] into rice that enabled the plants to manufacture beta-carotene in their endosperm.

2) Insect Resistance[6]

Bacillus thuringiensis[7] is a bacterium that is pathogenic[8] for a number of insect pests. Its lethal effect is mediated by a protein toxin it produces. Through recombinant DNA methods, the toxin gene can be introduced directly into the genome of the plant where it is expressed and provides protection against insect pests of the plant.

3) Disease Resistance[9]

Genes that provide resistance against plant viruses have been successfully introduced into such crop plants as tobacco, tomatoes, and potatoes.

4) Herbicide Resistance[10]

Questions have been raised about the safety—both to humans and to the environment—of some of the broad-leaved weed killers like 2,4-D. Alternatives are available, but they may damage the crop as well as the weeds growing in it. However, genes for resistance to some of the newer herbicides[11] have been introduced into some crop plants and enable them to thrive even when exposed to the weed killer.

5) Salt Tolerance[12]

A large fraction of the world's irrigated crop land is so laden with salt that it cannot be used to grow most important crops. However, researchers at the University of California Davis campus have created transgenic[13] tomatoes that grew well in saline soils. The transgene was a highly-expressed sodium/proton antiport pump[14] that sequestered excess sodium in the vacuole[15] of leaf cells. There was no sodium buildup in the fruit.

6) "Terminator" Genes[16]

This term is used (by opponents of the practice) for transgenes introduced into crop plants to make them produce sterile seeds (and thus force the farmer to buy fresh seeds for the following season rather than saving seeds from the current crop).

The process involves introducing three transgenes into the plant: ①A gene encoding a toxin which is lethal to developing seeds but not to mature seeds or the plant. This gene is normally inactive because of a stretch of DNA inserted between it and its promoter[17]. ②A gene encoding[18] a recombinase[19]—an enzyme that can remove the spacer in the toxin gene thus allowing to be expressed. ③A repressor gene whose protein product binds to the promoter of the recombinase thus keeping it inactive.

7) Transgenes Encoding Antisense RNA[20]

Messenger RNA[21] (mRNA) is single-stranded. Its sequence of nucleotides[22] is called "sense" because it results in a gene product (protein). Normally, its unpaired nucleotides are "read" by transfer RNA[23] anticodons[24] as the ribosome[25] proceeds to translate the message. However, RNA can form duplexes just as DNA does. All that is needed is a second strand of RNA whose sequence of bases is complementary[26] to the first strand. The second strand is called the antisense strand because its sequence of nucleotides is the complement of message sense. When mRNA forms a duplex with a complementary antisense RNA sequence, translation is blocked. This may occur because the ribosome cannot gain access to the nucleotides in the mRNA or duplex RNA[27] is quickly degraded by ribonucleases[28] in the cell. With recombinant DNA methods, synthetic genes (DNA) encoding antisense RNA molecules can be introduced into the organism.

8) Biopharmaceuticals[29]

Corn is the most popular plant for these purposes, but tobacco, tomatoes, potatoes, and rice are also being used. Some of the proteins that are being produced by transgenic crop plants: ①Human growth hormone with the gene inserted into the chloroplast DNA of tobacco plants. ②Humanized antibodies[30] against such infectious agents as (Ⅰ) HIV, (Ⅱ) respiratory syncytial virus[31] (RSV), (Ⅲ) sperm[32] (a possible contraceptive[33]), (Ⅳ) herpes simplex virus (HSV), the cause of "cold sores[34]". ③Protein antigens to be used in vaccines. An example: patient-specific anti-lymphoma[35] (a cancer) vaccines. B-cell lymphomas are clones of malignant B cells expressing on their surface a unique antibody molecule. Making tobacco plants transgenic for the RNA of the variable (unique) regions of this antibody enables them to produce the corresponding protein. This can then be incorporated into a vaccine in the hopes (early trials look promising) of boosting the patient's immune system—especially the cell-mediated branch—to combat the cancer. ④other useful proteins like lysozyme[36] and trypsin.

Notes to the Difficult Sentences

The process involves introducing three transgenes into the plant: ①A gene encoding a toxin which is lethal to developing seeds but not to mature seeds or the plant. This gene is normally inactive because of a stretch of DNA inserted between it and its promoter. ②A gene encoding a recombinase —an enzyme that can remove the spacer in the toxin gene thus allowing to be expressed. ③A repressor gene whose protein product binds to the promoter of the recombinase thus keeping it inactive.

该处理过程涉及将三个输入基因引入植物：①一个是编码对发育中的种子致命但对成熟种子和植株无害的毒素的基因。该基因通常情况下是无活性的，因为在其与其启动子之间插入了一段DNA。②一个是编码重组酶的基因，该酶能去除毒素基因间隔而使其表达。③一个是抑制子基因，其蛋白质产物与重组酶启动子结合可使其失活。

However, RNA can form duplexes just as DNA does. All that is needed is a second strand of RNA whose sequence of bases is complementary to the first strand. The second strand is called the antisense strand because its sequence of nucleotides is the complement of message sense. When mRNA forms a duplex with a complementary antisense RNA sequence, translation is blocked. This may occur because the ribosome cannot gain access to the nucleotides in the mRNA or duplex RNA is quickly degraded by ribonucleases in the cell.

然而，RNA 能像 DNA 一样形成双链。所需要的是 RNA 第二个链的碱基序列与第一个链互补。因为其核苷酸序列与正义链是互补的，所以第二个链称为反义链。mRNA 与互补的反义链形成双链时，翻译就被阻断。该现象的发生可能是因为在细胞内核糖体不能接近 mRNA 的核苷酸或双链 RNA 被核糖核酸酶快速降解。

Professional Words and Phrases

[1] **trait** [treɪt] n. 性状，品质
[2] **genetic manipulation** 遗传（基因）操作
[3] **cross-breeding** 杂交育种
[4] **catalyze** [ˈkætəlaɪz] v. 催化
[5] **transgene** [trænzˈdʒiːn] n. 输入基因
[6] **insect resistance** 抗虫性
[7] ***Bacillus thuringiensis*** 苏云金芽孢杆菌
[8] **pathogenic** [pæθəˈdʒenɪk] adj. 致病的
[9] **disease resistance** 抗病性
[10] **herbicide resistance** 除草剂抗性
[11] **herbicide** [ˈhəːbəsaɪd] n. 除草剂
[12] **salt tolerance** 耐盐性
[13] **transgenic** [trænzˈdʒenɪk] n. 转基因(做法), 转基因学; adj. 转基因的
[14] **sodium/proton antiport pump** 钠/氢逆向转运泵
[15] **vacuole** [ˈvækjuəul] n. 液泡
[16] **terminator gene** 终止子基因
[17] **promoter** [prəˈməutə] n. 启动子
[18] **encode** [ɪnˈkəud] vt. 编码
[19] **recombinase** [riːˈkɔmbəˈneɪs] n. 重组酶
[20] **antisense RNA** 反义 RNA
[21] **messenger RNA** 信使 RNA
[22] **nucleotide** [ˈnjuːklɪətaɪd] n. 核苷酸
[23] **transfer RNA** 转运 RNA
[24] **anticodon** [ˈæntɪˈkəudɔn] n. 反密码子
[25] **ribosome** [ˈraɪbəsəum] n. 核糖体
[26] **complementary** [kɔmplɪˈmentəri] adj. 互补的
[27] **duplex RNA** 双链 RNA
[28] **ribonuclease** [raɪbəuˈnjuːklɪeɪs] n. 核糖核酸酶
[29] **biopharmaceutical** [baɪəufɑːməˈsjuːtɪkəl] n. 生物制药; adj. 生物药剂学的
[30] **antibody** [ˈæntɪbɔdi] n. 抗体
[31] **respiratory syncytial virus** 呼吸道合胞病毒
[32] **sperm** [spəːm] n. 精子
[33] **contraceptive** [kɔntrəˈseptɪv] adj. 避孕的; n. 避孕药, 避孕用具

[34] **cold sores** 唇疱疹
[35] **lymphoma** [lim'fəumə] n. 淋巴癌
[36] **lysozyme** ['laisəzaim] n. 溶菌酶

Exercises

1. Matching

1) transgene   a) a type of pesticide used to kill unwanted plants
2) herbicide   b) a type of enzyme that can damage bacterial cell wall
3) promoter   c) a unit made up of three nucleotides that correspond to the three bases of the codon on the mRNA
4) antisense RNA   d) a gene or genetic material that has been transferred naturally or by any of a number of genetic engineering techniques from one organism to another
5) anticodon   e) a type of nuclease that catalyzes the degradation of RNA into smaller components
6) ribonuclease   f) a single-stranded RNA that is complementary to a messenger RNA (mRNA) strand transcribed within a cell
7) lysozyme   g) a regulatory region of DNA located upstream of a gene, providing a control point for regulated gene transcription

2. True or False

1) The synthesis of beta-carotene requires a number of enzyme-catalyzed steps.
2) *Bacillus thuringiensis* is a bacterium that is pathogenic for a number of mammalians including humans.
3) A repressor gene encodes a protein product binds to the promoter of the recombinase thus keeping it inactive.
4) When mRNA forms a duplex with a complementary antisense RNA sequence, translation is blocked.
5) The respiratory syncytial virus can lead to the "cold sores".

3. Reading Comprehension

1) Which crop is not mentioned to be related to transgenes against plant viruses?
   A. rice
   B. tobacco
   C. tomatoes
   D. potatoes

2) Which gene is not related to the production of sterile seeds?
   A. a gene encoding a recombinase—an enzyme that can remove the spacer in the toxin gene thus allowing to be expressed
   B. a gene encoding a highly-expressed sodium/proton antiport pump
   C. a gene encoding a toxin which is lethal to developing seeds but not to mature seeds or the plant
   D. a repressor gene whose protein product binds to the promoter of the recombinase thus

    keeping it inactive
3) Which crop is the most popular plant for biopharmaceuticals?
    A. corn
    B. tobacco
    C. tomatoes
    D. potatoes

4. Translation from English to Chinese

Bacillus thuringiensis is a bacterium that is pathogenic for a number of insect pests. Its lethal effect is mediated by a protein toxin it produces. Through recombinant DNA methods, the toxin gene can be introduced directly into the genome of the plant where it is expressed and provides protection against insect pests of the plant.

5. Translation from Chinese to English

世界上大部分灌溉农田由于充满了盐以至于不能种植最重要的农作物。然而，加利福尼亚大学戴维斯分校的研究人员已经研制出了在含盐土壤中能生长良好的转基因番茄。该输入基因是一个高表达的钠/氢逆向转运泵基因，它能阻止过量的钠进入叶肉细胞的液泡内。在果实中不会积累钠。

Chapter 5　Zoology—Animal Biology

[本章中文导读]

　　动物学是致力于研究动物的学科，是生物学的重要分支学科之一。本章主要介绍了什么是动物学（第5.1节），动物的组织、器官以及器官系统（第5.2节）和转基因动物（第5.3节）等与动物密切相关的知识点。作者想通过用英文讲解动物学一些基础科普知识，以及一些重要的动物学专业基础知识和当今动物学研究的热点技术——转基因动物等相关知识，达到使学生对动物学的专业英语词汇能够认识、理解、记忆的目的，同时通过介绍让学生感兴趣的动物学研究进展增强其学习兴趣。

5.1　What is Zoology?

　　Zoology is a branch of biology that focuses on the study of animals. Within this branch, people may also specialize and study certain forms of animals. For example a person working in zoology might study fish biology and work as an ichthyologist[1]. Alternately, a zoologist could specialize in the study of mammals and be called a mammalogist[2].

　　The discipline of zoology dates to the Ancient Greeks. In fact our primary system of classifying animals is largely based on Aristotle's work, though he made numerous classification errors. Early zoologists who followed Aristotle include many unknown collectors of animals who attempted to understand animals, or who were merely fascinated by the animal kingdom.

　　Modern zoology arrived with the invention of the microscope, but it would be a mistake to leave out the extremely important contributions of Charles Darwin. With Darwin's theory of evolution, understanding the relationship between humans and animals increased and further spurred interest in animals because humans were thought by some to have descended from them. As Darwin's theory gained acceptance, humans became part of the animal kingdom in a way that was both inclusive and humbling.

　　Zoologists that followed Darwin often studied animals in relationship to humans. Knowledge of the possible relationships of animals changed some classifications, and studies in zoology continue to refine animal classifications. The more zoologists learn about animals, the more diverse and complex the field of zoology becomes.

　　The field of zoology is a large one, with numerous specialists and numerous disciplines attached to it. Zoology scientists might study all animals, and branch out into ecological studies, or they may evaluate the chemistry of certain animals. Some zoologists devote their work to the study of one animal only, while other zoology scientists evaluate the lives of long extinct animal populations in the field called paleozoology[3].

　　A paleozoologist[4] might study with paleontologists, paleobotanists[5], and may include physical anthropology[6] in their studies. Anthrozoologists[7] study both cultural and physical anthropology since they principally study interactions between animals and people. One interesting branch of zoology is cryptozoology[8], which evaluates and attempts to classify animals only rumored to exist, like the Loch Ness monster or the Yeti. Cryptozoologists[9] might also do field studies if an animal thought previously extinct is re-discovered.

Zoology is certainly an important field, enhancing our understanding of the world in which we live. For the zoologist, there are numerous working opportunities, though some may not be particularly lucrative financially. Zoologists may contribute their studies to animal behaviorists, work in zoos, study animals in the wild, participate in archaeological[10] digs, apply their knowledge to improve raising of animals for food, be part of cloning experiments, or work primarily in labs analyzing the cell biology of animals. This large disciple, covering a variety of subfields, also may act as accessory to many related branches of science.

Notes to the Difficult Sentences

A paleozoologist might study with paleontologists, paleobotanists, and may include physical anthropology in their studies. Anthrozoologists study both cultural and physical anthropology since they principally study interactions between animals and people. One interesting branch of zoology is cryptozoology, which evaluates and attempts to classify animals only rumored to exist, like the Loch Ness monster or the Yeti.

古动物学家可能会与古生物学家和古植物学家一起进行研究，并且可能研究体质人类学。人文动物学者主要研究的是动物与人类的相互关系，他们不仅研究文化人类学而且研究体质人类学。一个非常有趣的动物学分支是神秘动物学，该学科评价并且试图将像尼斯湖水怪或西藏高原雪人等这样的仅仅是传说中存在的动物进行分类。

Professional Words and Phrases

[1] **ichthyologist** [ikθi'ɔlədʒist] n. 鱼类学者
[2] **mammalogist** [mæ'mælədʒist] n. 哺乳动物学者
[3] **paleozoology** [peiliəuzəu'ɔlədʒi] n. 古动物学
[4] **paleozoologist** [peiliəuzəu'ɔlədʒist] n. 古动物学家
[5] **paleobotanist** [peiliə'bɔtənist] n. 古植物学家
[6] **physical anthropology**   体质人类学
[7] **anthrozoologist** [ænθrəzəu'ɔlədʒist] n. 人文动物学者
[8] **cryptozoology** ['kriptəuzəu'ɔlədʒi] n. 神秘动物学
[9] **cryptozoologist** ['kriptəuzəu'ɔlədʒist] n. 神秘动物学研究者
[10] **archaeological** [aːkiə'lɔdʒikəl] adj. 考古学的

Exercises

1. Matching

1) ichthyologist   a) a zoologist specializing in the study of mammals
2) mammalogist   b) a zoologist focusing on the study of paleozoology

3) paleozoologist   c) a person studying fish biology

2. True or False

1) A zoologist could specialize in the study of mammals and be called a mammalogist.

2) Humans are still accepted to have descended from them now.

3) Cryptozoology is used for the evaluation and classification of animals only rumored to exist, like the Loch Ness monster or the Yeti.

4) Zoology scientists might study all animals, and branch out into ecological studies, or they may evaluate the chemistry of certain animals.

5) Charles Darwin made the extremely important contributions to the modern zoology.

3. Reading Comprehension

1) Which is not included in zoologists?
   A. an ichthyologist
   B. a mammalogist
   C. a paleobotanist
   D. a paleozoologist

2) Which is not the branch of zoology?
   A. fish biology
   B. microbiology
   C. anthropology
   D. cryptozoology

3) Which of the following statements is not true?
   A. Aristotle made the contributions to the early zoology, though he made numerous classification errors.
   B. Cryptozoologists do not study an animal thought previously extinct is re-discovered.
   C. A paleozoologist evaluate the lives of long extinct animal populations.
   D. Anthrozoologists principally study interactions between animals and people.

4. Translation from English to Chinese

　　Modern zoology arrived with the invention of the microscope, but it would be a mistake to leave out the extremely important contributions of Charles Darwin. With Darwin's theory of evolution, understanding the relationship between humans and animals increased and further spurred interest in animals because humans were thought by some to have descended from them. As Darwin's theory gained acceptance, humans became part of the animal kingdom in a way that was both inclusive and humbling.

5. Translation from Chinese to English

　　动物学是致力于研究动物的一个生物学分支。在这一分支中，人们也可以只专一地研究某一类动物。例如，在动物学领域工作的人可以只作为一名鱼类学者研究鱼类生物学。再如，动物学家也能只专一地研究哺乳动物，而被称为哺乳动物学者。

5.2　Tissues, Organs and Organ Systems of Animals

There are functions which every animal and organism, from the simplest to the most complex, must perform. These include the intake of nutrients (digestion[1] in us), respiration[2] or gas exchange, excretion[3] (ridding the body of toxic wastes), coordination of actions (nerves[4] and glands[5]), movement (use of muscles in animals) and reproduction. These functions are carried out by organs and/or organ systems[6]. The organs, in turn, are composed of a variety of tissues. Tissues are associations of similar cells which carry out a specialized function. In general, vertebrates have the same organ systems and tissues.

Tissues

The basic unit of life is the cell, and the cells of complex organisms are organized into tissues. Both plants and animals have tissues and organs but we will focus on the tissues and organs of animals, the vertebrates[7] specifically. In vertebrates, tissues are derived from the three layers of the embryo: the ectoderm[8] (outer layer) gives rise to the skin and tissues of the nervous system[9]; the mesoderm[10] (middle layer) gives rise to muscle, bone, and many of the reproductive, urinary[11] and circulatory organs; and the endoderm[12] (inner layer) gives rise to the lining of the digestive tract[13] and organs derived from it such as the lungs. Tissues are composed of cells and extracellular products. Organs are composed of several types of tissues and organ systems are composed of several organs. The organ systems carry out the functions of the body such as digestion, communication, circulation, respiration, excretion, and movement.

Tissues may be categorized into four major types: ①epithelial[14], ②connective tissue[15], ③muscle, and ④nervous tissue. The cells of tissues are held together by one or more of a variety of cell junctions[16]. Some are tight junctions which do not let fluids pass, some are composed of supporting filaments to give the cell shape and to attach the cell to its neighboring cells (adhering junctions), and some, such as the gap junctions[17], are for intercellular communication.

Organs and Organ Systems

Organs are functional units of an animal's body that are made up of more than one type of tissue. Examples include the heart, lungs, liver, spleen[18], and kidneys[19].

The next higher level of structural organization in animals is the organ system. An organ system is an association of organs that together perform an overall function. The major organ systems of the body and their functions are:

1) integumentary system[20] for protection, excretion, receipt of external stimuli (outer covering of skin).

2) muscular system[21] for movement, posture, heat production.

3) skeletal system for support, muscle attachment, red blood cell production, Ca^{2+} and PO_4^{3-} storage.

4) nervous system[22] to rapidly detect external and internal stimuli and coordinate responses.

5) endocrine system[23] which produces hormones to complement the nervous system in the control of body functions. Hormonal coordination which is coordination by biochemicals secreted and detected by specialized cells is the most ancient form of intercellular communication. It is used by every type of organism including the Monera[24].

6) circulatory system[25] functions to deliver biomolecules[26] (including gases) to and pick up waste or toxic biomolecules from all cells of the organism. In vertebrates it is a "closed" system and works in close conjunction with the respiratory system[27].

7) The lymphatic system[28] works in conjunction with the circulatory system to assist in picking up excess fluid (lymph[29]) released by the circulatory system (capillaries[30]) into the tissues. It is an "open" system which begins in the tissues not with arteries[31] and arterioles[32]. When lymph vessels[33] are blocked by parasites[34] such as the roundworm[35], *Wuchereria bancrofti*[36], a condition known as elephantiasis[37] results. As the name implies, the portion of the body affected reaches elephantine size.

8) The respiratory system functions in the exchange of the gases, oxygen and carbon dioxide. In vertebrates the blood vessels[38] of the lungs bring deoxygenated[39] blood into the lung and carry away oxygenated blood. Carbon dioxide is downloaded in the lung as well.

9) The digestive system[40] functions to process nutrients for energy and as building blocks for the body. Large pieces of food are systematically broken down to smaller units and eventually chemically broken down into biomolecules which can be absorbed in the gut and into the circulatory system. The undigested remains are eliminated.

10) The urinary system[41] has the job of maintaining a constant internal environment. Through it passes the blood for filtering and maintenance of osmotic[42] equilibrium[43], the retention of valuable nutrients and the excretion of excess or toxic molecules.

11) The reproductive system[44] is differentiated into male and female and consists of gonads[45] and appropriate internal and external "plumbing" or ducts. Its job is to produce gametes (eggs or sperm) and to deliver them to the appropriate site for fertilization[46]. Depending on whether the organism has internal or external fertilization or internal or external development of the embryo, additional organs are added as necessary. The gonads produce both gametes (exocrine[47] function) and hormones (endocrine[48] function). The sex hormones influence other organs as well (e.g., skin, muscle, breast, uterus[49], phallus[50], etc.)

Notes to the Difficult Sentences

The reproductive system is differentiated into male and female and consists of gonads and appropriate internal and external "plumbing" or ducts. Its job is to produce gametes (eggs or sperm) and to deliver them to the appropriate site for fertilization. Depending on whether the organism has internal or external fertilization or internal or external development of the embryo, additional organs are added as necessary. The gonads produce both gametes (exocrine function) and hormones (endocrine function). The sex hormones influence other organs as well (e.g., skin, muscle, breast, uterus, phallus, etc.)

生殖系统分化成雄性和雌性两种，并且由腺体和适当的内管道和外管道组成。其功能是产生配子（卵子和精子）并将它们运到合适的位置进行受精。生命体会视其是否需要进行内部或外部受精，或在胚胎内发育或胚胎外发育，在必要时增加其他器官。腺体的功能是产生配子（外分泌功能）和激素（内分泌功能）。性激素同时也影响其他器官的发育（例如皮肤、肌肉、乳房、子宫和阴茎等）。

Professional Words and Phrases

[1] **digestion** [di'dʒestʃən] n. 消化，吸收
[2] **respiration** [respə'reiʃən] n. 呼吸
[3] **excretion** [eks'kri:ʃən] n. 排泄
[4] **nerve** [nə:v] n. 神经
[5] **gland** [glænd] n. 腺
[6] **organ system** 器官系统
[7] **vertebrate** ['və:tibreit] n. 脊椎动物
[8] **ectoderm** ['ektəudə:m] n. 外胚层
[9] **nervous system** 神经系统
[10] **mesoderm** ['mesədə:m] n. 中胚层
[11] **urinary** ['juərəneri] adj. 泌尿（器官）的，尿的
[12] **endoderm** ['endəudə:m] n. 内胚层
[13] **digestive tract** 消化道
[14] **epithelial** [epi'θi:liəl] adj. 上皮的
[15] **connective tissue** 结缔组织
[16] **cell junction** 细胞连接
[17] **gap junction** 缝隙连接
[18] **spleen** [spli:n] n. 脾
[19] **kidney** ['kidni] n. 肾
[20] **integumentary system** 皮肤系统
[21] **muscular system** 肌肉系统
[22] **nervous system** 神经系统
[23] **endocrine system** 内分泌系统
[24] **Monera** [mə'niərə] n. 无核原虫界
[25] **circulatory system** 循环系统
[26] **biomolecule** [baiəu'mɔlikju:l] n. 生物分子
[27] **respiratory system** 呼吸系统
[28] **lymphatic system** 淋巴系统
[29] **lymph** [limf] n. 淋巴，淋巴液
[30] **capillary** ['kæpiləri] n. 毛细管; adj. 毛细管的
[31] **artery** ['ɑ:təri] n. 动脉
[32] **arteriole** [ɑ:'tiəriəul] n. 细动脉
[33] **lymph vessel** 淋巴管
[34] **parasite** ['pærəsait] n. 寄生虫，寄生生物
[35] **roundworm** ['raundwə:m] n. 蛔虫
[36] *Wuchereria bancrofti* 班克罗夫特线虫
[37] **elephantiasis** [eləfən'taiəsis] n. 象皮病
[38] **blood vessel** 血管
[39] **deoxygenate** [di:'ɔksədʒəneit] vt. 除去氧气

[40] **digestive system** 消化系统
[41] **urinary system** 泌尿系统
[42] **osmotic** [ɔz'mɔtik] adj. 渗透的
[43] **equilibrium** [iːkwi'libriəm] n. 平衡
[44] **reproductive system** 生殖系统
[45] **gonad** ['gɔnæd] n. 性腺
[46] **fertilization** [fəːtilai'zeiʃən] n. 受精, 肥沃化
[47] **exocrine** ['eksəkrin] adj. 外分泌的
[48] **endocrine** ['endəukrain] adj. 内分泌的
[49] **uterus** ['juːtərəs] n. 子宫
[50] **phallus** ['fæləs] n. 阴茎

Exercises

1. Matching

1) excretion　　a) a muscular organ of the female mammal for containing and usually for nourishing the young during development prior to birth

2) vertebrate　　b) a disease that is characterized by the thickening of the skin and underlying tissues, especially in the legs and male genitals

3) endocrine system　　c) any of the small terminal twigs of an artery that ends in capillaries

4) lymphatic system　　d) an act or process of making fertile

5) elephantiasis　　e) a organ system of glands, each of which secretes a type of hormone directly into the bloodstream to regulate the body

6) fertilization　　f) the act or process of excreting

7) uterus　　g) the part of the immune system comprising a network of conduits called lymphatic vessels

8) arteriole　　h) any of a subphylum (Vertebrata) of chordates possessing a spinal column that includes the mammals, birds, reptiles, amphibians, and fishes

2. True or False

1) Tissues are associations of similar cells which carry out a specialized function.
2) Integumentary system is used for movement, posture, heat production.
3) The urinary system has the job of maintaining a constant internal environment.
4) The respiratory system functions to process nutrients for energy and as building blocks for the body.
5) The circulatory system functions to deliver biomolecules (including gases) to and pick up waste or toxic biomolecules from all cells of the organism.

3. Reading Comprehension

1) Which is the basic unit of life?
　A. cell
　B. tissue
　C. organs

D. organ systems

2) Which organ system is used for support, muscle attachment, red blood cell production, Ca^{2+} and PO_4^{3-} storage?

   A. urinary system

   B. digestive system

   C. reproductive system

   D. skeletal system

3) Which layer of the embryo gives rise to the lungs?

   A. the ectoderm

   B. the mesoderm

   C. the endoderm

   D. None of the above

4. Translation from English to Chinese

In vertebrates, tissues are derived from the three layers of the embryo: the ectoderm (outer layer) gives rise to the skin and tissues of the nervous system; the mesoderm (middle layer) gives rise to muscle, bone, and many of the reproductive, urinary and circulatory organs; and the endoderm (inner layer) gives rise to the lining of the digestive tract and organs derived from it such as the lungs. Tissues are composed of cells and extracellular products. Organs are composed of several types of tissues and organ systems are composed of several organs. The organ systems carry out the functions of the body such as digestion, communication, circulation, respiration, excretion, and movement.

5. Translation from Chinese to English

组织可分为四种主要类型：①表皮组织，②结缔组织，③肌肉组织，以及④神经组织。组织细胞是由一个或数个类型不同的细胞连接组合到一起形成的。有些细胞连接是紧密连接，是不让液体通过的；有些则由支架细丝组成的，用于维持细胞形态并将细胞贴附在相邻细胞上（黏着连接），而有些像缝隙连接这样的连接则是用于细胞间交流的。

5.3　Transgenic Animals—Additional Reading

Nowadays, breakthroughs in molecular biology are happening at an unprecedented rate. One of them is the ability to engineer transgenic animals[1], i.e., animals that carry genes from other species. The technology has already produced transgenic animals such as mice, rats, rabbits, pigs, sheep, and cows.

What is a transgenic animal?

There are various definitions for the term transgenic animal. The Federation of European Laboratory Animal Associations defines the term as an animal in which there has been a deliberate modification of its genome, the genetic makeup[2] of an organism responsible for inherited characteristics.

The nucleus of all cells in every living organism contains genes made up of DNA. These genes store information that regulates how our bodies form and function. Genes can be altered artificially, so that some characteristics of an animal are changed. For example, an embryo can have an extra, functioning gene from another source artificially introduced into, or a gene introduced which can knock out[3] the functioning of another particular gene in the embryo. Animals that have their DNA manipulated in this way are known as transgenic animals.

The majority of transgenic animals produced so far are mice, the animal that pioneered the technology. The first successful transgenic animal was a mouse. A few years later, it was followed by rabbits, pigs, sheep, and cattle.

Why are these animals being produced?

The two most common reasons are: ①Some transgenic animals are produced for specific economic traits. For example, transgenic cattle were created to produce milk containing particular human proteins, which may help in the treatment of human emphysema[4]. ②Other transgenic animals are produced as disease models (animals genetically manipulated to exhibit disease symptoms so that effective treatment can be studied). For example, Harvard scientists made a major scientific breakthrough when they received a U.S. patent (the company DuPont holds exclusive rights to its use) for a genetically engineered mouse, called OncoMouse® or the Harvard mouse, carrying a gene that promotes the development of various human cancers.

How are transgenic animals produced?

Since the discovery of the molecular structure of DNA by Watson and Crick in 1953, molecular biology research has gained momentum. Molecular biology technology combines techniques and expertise from biochemistry, genetics, cell biology, developmental biology[5], and microbiology. Scientists can now produce transgenic animals because, since Watson and Crick's discovery, there have been breakthroughs in: ①recombinant DNA (artificially-produced DNA); ②genetic cloning[6]; ③analysis of gene expression (the process by which a gene gives rise to a protein); ④genomic mapping[7].

The underlying principle in the production of transgenic animals is the introduction of a foreign gene[8] or genes into an animal (the inserted genes are called transgenes). The foreign genes "must be transmitted through the germ line[9], so that every cell, including germ cells[10], of the animal contains the same modified genetic material." (Germ cells are cells whose function is to transmit genes to an organism's offspring.) To date, there are three basic methods of producing transgenic animals: ①DNA microinjection[11], ②Retrovirus[12]-mediated gene transfer[13], ③Embryonic[14] stem cell[15]-mediated gene transfer.

Gene transfer by microinjection is the predominant method used to produce transgenic farm animals. Since the insertion of DNA results in a random process, transgenic animals are mated to ensure that their offspring acquire the desired transgene. However, the success rate of producing transgenic animals individually by these methods is very low and it may be more efficient to use cloning techniques to increase their numbers. For example, gene transfer studies revealed that only 0.6% of transgenic pigs were born with a desired gene after 7,000 eggs were injected with a specific transgene.

Notes to the Difficult Sentences

The nucleus of all cells in every living organism contains genes made up of DNA. These genes store information that regulates how our bodies form and function. Genes can be altered artificially, so that some characteristics of an animal are changed. For example, an embryo can have an extra, functioning gene from another source artificially introduced into, or a gene introduced which can knock out the functioning of another particular gene in the embryo. Animals that have their DNA manipulated in this way are known as transgenic animals.

每个生命体中所有细胞的核均含有由DNA组成的基因。这些基因储存着调控机体形成和行使功能的信息。基因可以被人工改造，这样动物的一些特征就可以被改变。例如，一个胚胎能含有一个额外的、从其他来源引入的、有功能的基因，或是一个能敲除胚胎中另一特定基因的输入基因。具有以该方式操作过的DNA的动物被称作转基因动物。

Professional Words and Phrases

[1]　**transgenic animal**　转基因动物
[2]　**genetic makeup**　基因组合
[3]　**knock out**　敲除（基因）
[4]　**emphysema**　[emfi'si:mə]　n. 肺气肿
[5]　**developmental biology**　发育生物学
[6]　**genetic cloning**　遗传（基因）克隆
[7]　**genomic mapping**　基因组作图
[8]　**foreign gene**　外源基因
[9]　**germ line**　种系
[10]　**germ cell**　生殖细胞，受精卵
[11]　**microinjection**　[maikrəuin'dʒekʃən]　n. 显微注射
[12]　**retrovirus**　[retrəu'vaiərəs]　n. 逆转录病毒
[13]　**gene transfer**　基因转移
[14]　**embryonic**　[embri'ɔnik]　adj. 胚胎的
[15]　**stem cell**　干细胞

Exercises

1. Matching

1) transgenic animal　　a) the creation of a genetic map assigning DNA fragments to chromosomes

2) emphysema　　b) the process of using a glass micropipette to insert substances at a microscopic or borderline macroscopic level into a single living cell

3) genetic cloning　　c) biological cells found in all multicellular organisms, that can divide through mitosis and differentiate into diverse specialized cell types

4) genomic mapping　　d) an animal that carry genes from other species

5) microinjection　　e) the replication of DNA fragments by the use of a self-replicating genetic material

6) gene transfer　　f) a long-term, progressive disease of the lungs that primarily causes

shortness of breath

7) stem cell   g) foreign genetic material, either DNA or RNA, is introduced artificially or naturally into a cell

2. True or False

1) Current breakthroughs in molecular biology are happening at an unprecedented rate.

2) Disease models refers to animals genetically manipulated to exhibit disease symptoms so that effective treatment can be studied.

3) Germ cells are cells whose function is to transmit genes to an organism's offspring.

3. Reading Comprehension

1) Which one is the first successful transgenic animal?

   A. a mouse

   B. a rabbit

   C. a cattle

   D. a pig

2) Which method is the predominant method used to produce transgenic farm animals?

   A. DNA microinjection

   B. Retrovirus-mediated gene transfer

   C. Embryonic stem cell-mediated gene transfer

   D. None of the above

3) Which of the following statements is not true?

   A. A gene can be introduced to knock out the functioning of another particular gene in the embryo.

   B. Gene expression means the process by which a gene gives rise to a protein.

   C. Gene transfer by microinjection is the predominant method for transgenic animals just because of its success rates.

   D. The nucleus of all cells in every living organism contains genes made up of DNA.

4. Translation from English to Chinese

The underlying principle in the production of transgenic animals is the introduction of a foreign gene or genes into an animal (the inserted genes are called transgenes). The foreign genes "must be transmitted through the germ line, so that every cell, including germ cells, of the animal contain the same modified genetic material." (Germ cells are cells whose function is to transmit genes to an organism's offspring.) To date, there are three basic methods of producing transgenic animals: ①DNA microinjection, ②Retrovirus-mediated gene transfer, ③Embryonic stem cell-mediated gene transfer.

5. Translation from Chinese to English

借助显微注射的基因转移是产生农场转基因动物的最主要方法。由于DNA的插入是一个随机的过程，转基因动物通过交配才能确保在它们后代中获得人们想要的输入基因。然而，单一地通过这些方法产生转基因动物的成功率是很低的，而使用克隆技术可能会更有效地提高转基因动物的数量。

Chapter 6　Molecular Genetics

[本章中文导读]

　　分子遗传学作为当今最热门的学科，已经渗透到生物学的各个分支学科，甚至是医学、环境科学、食品科学等其他的相关学科。该领域的专业英语知识的学习对于生物学专业学生具有至关重要的意义。本章主要介绍了基因和染色体（第6.1节）、基因的功能结构（第6.2节）、基因表达（第6.3节）以及基因工程的必要条件（第6.4节）等分子遗传学方面重要的知识点。作者想通过用英文讲解分子遗传学的专业基础知识以及由分子遗传学衍生出来的基因工程技术等相关知识，达到使学生初步掌握分子遗传学的专业英语词汇的具体含义以及使用方法的目的。

6.1　Gene and Chromosomes

Gene

A gene is a unit of heredity[1] in a living organism. It is a name given to some stretches of DNA and RNA that code for a type of protein or for an RNA chain that has a function in the organism. Living things depend on genes, as they specify all proteins and functional RNA chains. Genes hold the information to build and maintain an organism's cells and pass genetic traits[2] to offspring, although some organelles (e.g. mitochondria[3]) are self-replicating and are not coded for by the organism's DNA. All organisms have many genes corresponding to many different biological traits[4], some of which are immediately visible, such as eye color or number of limbs, and some of which are not, such as blood type or increased risk for specific diseases, or the thousands of basic biochemical processes that comprise life.

A modern working definition of a gene is "a locatable region of genomic sequence, corresponding to a unit of inheritance[5], which is associated with regulatory regions, transcribed regions, and or other functional sequence regions". Colloquial usage of the term gene (e.g. "good genes", "hair color gene") may actually refer to an allele[6]: a gene is the basic instruction, a sequence of nucleic acids (DNA or, in the case of certain viruses RNA), while an allele is one variant[7] of that gene. Thus, when the mainstream press refers to "having" a "gene" for a specific trait, this is generally inaccurate. In most cases, all people would have a gene for the trait in question, but certain people will have a specific allele of that gene, which results in the trait variant. In the simplest case, the phenotypic[8] variation[9] observed may be caused by a single letter of the genetic code—a single nucleotide polymorphism[10].

Chromosomes

The total complement of genes in an organism or cell is known as its genome, which may be stored on one or more chromosomes; the region of the chromosome at which a particular gene is located is called its locus. A chromosome consists of a single, very long DNA helix on which thousands of genes are encoded.

Prokaryotes—bacteria and archaea[11]—typically store their genomes on a single large, circular chromosome, sometimes supplemented by additional small circles of DNA called plasmids[12], which usually encode only a few genes and are easily transferable between individuals. For example, the genes for antibiotic resistance are usually encoded on bacterial plasmids and can be passed between individual cells, even those of different species, via horizontal gene transfer[13].

Although some simple eukaryotes also possess plasmids with small numbers of genes, the majority of eukaryotic genes are stored on multiple linear chromosomes, which are packed within the nucleus in complex with storage proteins called histones[14]. The manner in which DNA is stored on the histone, as well as chemical modifications of the histone itself, are regulatory mechanisms governing whether a particular region of DNA is accessible for gene expression. The ends of eukaryotic chromosomes are capped by long stretches of repetitive sequences called telomeres[15], which do not code for any gene product but are present to prevent degradation of coding and regulatory regions during DNA replication. The length of the telomeres tends to decrease each time the genome is replicated in preparation for cell division; the loss of telomeres has been proposed as an explanation for cellular senescence[16], or the loss of the ability to divide, and by extension for the aging process in organisms.

Whereas the chromosomes of prokaryotes are relatively gene-dense, those of eukaryotes often contain so-called "junk DNA[17]", or regions of DNA that serve no obvious function. Simple single-celled eukaryotes have relatively small amounts of such DNA, whereas the genomes of complex multicellular organisms, including humans, contain an absolute majority of DNA without an identified function. However it now appears that, although protein-coding DNA makes up barely 2% of the human genome, about 80% of the bases in the genome may be expressed, so the term "junk DNA" may be a misnomer.

Notes to the Difficult Sentences

Colloquial usage of the term gene (e.g. "good genes", "hair color gene") may actually refer to an allele: a gene is the basic instruction, a sequence of nucleic acids (DNA or, in the case of certain viruses RNA), while an allele is one variant of that gene. Thus, when the mainstream press refers to "having" a "gene" for a specific trait, this is generally inaccurate. In most cases, all people would have a gene for the trait in question, but certain people will have a specific allele of that gene, which results in the trait variant. In the simplest case, the phenotypic variation observed may be caused by a single letter of the genetic code—a single nucleotide polymorphism.

专业术语"基因"（如好的基因、发色基因）的口头上的使用事实上可能指的是等位基因：基因是基本指令，是核酸序列（DNA或是在特定病毒中的RNA），而等位基因是基因的变体。因此，当主流媒体提到某个负责特定性状的基因，通常是不准确的。多数情况下，任何人都具有人们所提到的负责某一特定性状的基因，但是只有某些人具有能导致该性状变异的特定等位基因。最简单的一个例子是人们所观察到的表型变化可能是由遗传密码的单个字母——单核苷酸多态性导致的。

Professional Words and Phrases

[1] **heredity** [hiˈrediti] n. 遗传
[2] **genetic trait** 遗传性状
[3] **mitochondria** [maitəˈkɔndriə] n. 线粒体, mitochondrium 的复数形式
[4] **biological trait** 生物性状
[5] **inheritance** [inˈheritəns] n. 遗传，继承物
[6] **allele** [əˈliːl] n. 等位基因
[7] **variant** [ˈveriənt] n. 变体，变型
[8] **phenotypic** [ˈfiːnətipik] adj. 表型的
[9] **variation** [veriˈeiʃən] n. 变种，变化
[10] **polymorphism** [pɔliˈmɔːfizəm] n. 多态性
[11] **archaea** [ɑːˈkiə] n. 古菌
[12] **plasmid** [ˈplæzmid] n. 质粒
[13] **horizontal gene transfer** 水平基因转移
[14] **histone** [ˈhistəun] n. 组蛋白
[15] **telomere** [ˈteləmiə] n. 端粒
[16] **cellular senescence** 细胞衰老
[17] **junk DNA** 垃圾 DNA

Exercises

1. Matching

1) allele   a) the location of a gene in the genome
2) telomere   b) noncoding DNA describes components of an organism's DNA sequences that do not encode for protein sequences
3) plasmid   c) highly alkaline proteins found in eukaryotic cell nuclei that package and order the DNA into structural units called nucleosomes
4) histone   d) long stretches of repetitive sequences in the eukaryotic chromosome
5) junk DNA   e) a DNA molecule that is separate from, and can replicate independently of, the chromosomal DNA
6) locus   f) one of two or more forms of a gene

2. True or False

1) A gene refers to some stretches of DNA and RNA that code for a type of protein or for an RNA chain that has a function in the organism.
2) In most cases, all people would have an allele for the trait in question.
3) Prokaryotes typically store all of their genomes on a single large, circular chromosome.
4) The genes for antibiotic resistance are usually encoded on bacterial chromosomes.
5) The chromosomes of prokaryotes often contain so-called "junk DNA", or regions of DNA that serve no obvious function.
6) The ends of eukaryotic chromosomes are capped by long stretches of repetitive sequences

called telomeres.

7) The genomes of complex multicellular organisms still contain an absolute majority of DNA without an identified function.

3. Reading Comprehension

1) Which is a false description of a gene?

   A. A gene can pass genetic traits to offspring.

   B. A gene is a locatable region of genomic sequence.

   C. A gene is a stretch of DNA sequence in all cases.

   D. A gene code for a type of protein or for an RNA chain.

2) Which of the following statements is not true?

   A. A gene is a sequence of nucleic acids while an allele is one variant of that gene.

   B. DNA sequence of the telomeres is a good example of "junk DNA".

   C. Protein-coding DNA only makes up barely 2% of the human genome.

   D. The majority of eukaryotic genes are stored on multiple linear chromosomes.

3) What is the name of the gene location?

   A. allele

   B. locus

   C. plasmid

   D. telomere

4. Translation from English to Chinese

Prokaryotes—bacteria and archaea—typically store their genomes on a single large, circular chromosome, sometimes supplemented by additional small circles of DNA called plasmids, which usually encode only a few genes and are easily transferable between individuals. For example, the genes for antibiotic resistance are usually encoded on bacterial plasmids and can be passed between individual cells, even those of different species, via horizontal gene transfer.

5. Translation from Chinese to English

原核生物染色体上的基因密度是相当大的，而真核生物染色体经常含有所谓的"垃圾DNA"或没有明显功能的DNA区。在简单的单细胞真核生物中，这样的DNA的数量是相当小的，而包括人在内的复杂的多细胞生物的基因组含有的DNA绝大多数是尚未鉴定功能的。然而，现在看来尽管编码蛋白质的DNA仅占人类基因组的2%还不到，但是大约80%的基因组碱基均可以表达，所以说专业术语"垃圾DNA"可能是使用不当的名称。

6.2 Functional Structure of a Gene

The vast majority of living organisms encode their genes in long strands of DNA. DNA (deoxyribonucleic acid[1]) consists of a chain made from four types of nucleotide subunits[2], each composed of: a five-carbon sugar[3] (2'-deoxyribose[4]), a phosphate[5] group, and one of the four bases adenine[6], cytosine[7], guanine[8], and thymine[9]. The most common form of DNA in a cell is in a double helix structure[10], in which two individual DNA strands twist around each other in a right-handed spiral. In this structure, the base pairing[11] rules specify that guanine pairs with cytosine and adenine pairs with thymine. The base pairing between guanine and cytosine forms three hydrogen bonds[12], whereas the base pairing between adenine and thymine forms two hydrogen bonds. The two strands in a double helix must therefore be complementary, that is, their bases must align such that the adenines of one strand are paired with the thymines of the other strand, and so on.

Due to the chemical composition of the pentose[13] residues of the bases, DNA strands have directionality. One end of a DNA polymer contains an exposed hydroxyl group[14] on the deoxyribose; this is known as the 3' end of the molecule. The other end contains an exposed phosphate group; this is the 5' end. The directionality of DNA is vitally important to many cellular processes, since double helices are necessarily directional (a strand running 5'-3' pairs with a complementary strand running 3'-5'), and processes such as DNA replication[15] occur in only one direction. All nucleic acid synthesis in a cell occurs in the 5'-3' direction, because new monomers are added via a dehydration[16] reaction[17] that uses the exposed 3' hydroxyl as a nucleophile[18].

The expression[19] of genes encoded in DNA begins by transcribing[20] the gene into RNA, a second type of nucleic acid that is very similar to DNA, but whose monomers contain the sugar ribose[21] rather than deoxyribose. RNA also contains the base uracil[22] in place of thymine. RNA molecules are less stable than DNA and are typically single-stranded. Genes that encode proteins are composed of a series of three-nucleotide sequences called codons, which serve as the words in the genetic language. The genetic code specifies the correspondence during protein translation between codons and amino acids. The genetic code is nearly the same for all known organisms.

All genes have regulatory regions in addition to regions that explicitly code for a protein or RNA product. A regulatory region[23] shared by almost all genes is known as the promoter, which provides a position that is recognized by the transcription machinery[24] when a gene is about to be transcribed and expressed[25]. A gene can have more than one promoter, resulting in RNAs that differ in how far they extend in the 5' end. Although promoter regions have a consensus sequence[26] that is the most common sequence at this position, some genes have "strong" promoters[27] that bind the transcription machinery well, and others have "weak" promoters[28] that bind poorly. These weak promoters usually permit a lower rate of transcription than the strong promoters, because the transcription machinery binds to them and initiates transcription[29] less frequently. Other possible regulatory regions include enhancers[30], which can compensate for a weak promoter. Most regulatory regions are "upstream"—that is, before or toward the 5' end of the transcription initiation site[31]. Eukaryotic promoter regions are much more complex and difficult to identify than prokaryotic promoters.

Many prokaryotic genes are organized into operons[32], or groups of genes whose products have related functions and which are transcribed as a unit. By contrast, eukaryotic genes are transcribed only one at a time, but may include long stretches of DNA called introns[33] which are transcribed but never translated[34] into protein (they are spliced[35] out before translation[36]). Splicing[37] can also occur in prokaryotic genes, but is less common than in eukaryotes.

Notes to the Difficult Sentences

One end of a DNA polymer contains an exposed hydroxyl group on the deoxyribose; this is known as the 3' end of the molecule. The other end contains an exposed phosphate group; this is the 5' end. The directionality of DNA is vitally important to many cellular processes, since double helices are necessarily directional (a strand running 5'-3' pairs with a complementary strand running 3'-5'), and processes such as DNA replication occur in only one direction. All nucleic acid synthesis in a cell occurs in the 5'-3' direction, because new monomers are added via a dehydration reaction that uses the exposed 3' hydroxyl as a nucleophile.

DNA 聚合体一端含有一个裸露的脱氧核糖的羟基，这一端被称作 DNA 的 3'端。另一端含有一个裸露的磷酸基团，这一端被称为 DNA 的 5'端。DNA 双螺旋必须有方向性（即一条 5'-3'方向的链与一条 3'-5'方向的互补链配对），并且 DNA 复制过程仅从一个方向上发生，因此 DNA 的方向性对许多细胞过程是非常重要的。因为新的单体需要借助 3'端羟基作为亲核试剂通过脱水反应添加，所以细胞中所有核酸的合成均必须按 5'-3'方向进行。

Professional Words and Phrases

[1] **deoxyribonucleic acid**　脱氧核糖核酸
[2] **subunit**　[sʌb'juːnit]　n. 亚基，亚单位
[3] **five-carbon sugar**　五碳糖
[4] **deoxyribose**　[diːˌɔksi'raibəus]　n. 脱氧核糖
[5] **phosphate**　['fɔsfeit]　n. 磷酸盐
[6] **adenine**　['ædənin]　n. 腺嘌呤
[7] **cytosine**　['saitəusiːn]　n. 胞嘧啶
[8] **guanine**　['gwɑːniːn]　n. 鸟嘌呤
[9] **thymine**　['θaimiːn]　n. 胸腺嘧啶
[10] **double helix structure**　双螺旋结构
[11] **base pairing**　碱基配对
[12] **hydrogen bond**　氢键
[13] **pentose**　['pentəus]　n. 戊糖
[14] **hydroxyl group**　羟基
[15] **replication**　[repli'keiʃən]　n. 复制
[16] **dehydration**　[diːhai'dreiʃən]　n. 脱水
[17] **dehydration reaction**　脱水反应
[18] **nucleophile**　['njuːkliːəfail]　n. 亲核试剂
[19] **expression**　[iks'preʃən]　n.（基因）表达
[20] **transcribe**　[træns'kraib]　v. 转录
[21] **ribose**　['raibəus]　n. 核糖
[22] **uracil**　['juərəsil]　n. 尿嘧啶
[23] **regulatory region**　调控区
[24] **transcription machinery**　转录复合物
[25] **express**　[iks'pres]　vt.（基因）使表达

[26] **consensus sequence** 共有序列
[27] **strong promoter** 强启动子
[28] **weak promoter** 弱启动子
[29] **transcription** [træns'kripʃən] n. 转录
[30] **enhancer** [in'hɑːnsə] n. 增强子
[31] **transcription initiation site** 转录起始位点
[32] **operon** ['ɔpə,rɔn] n. 操纵子
[33] **intron** ['intrɔn] n. 内含子
[34] **translate** [træns'leit] v. （RNA）翻译
[35] **splice** [splais] vt. 剪切，拼接
[36] **translation** [træns'leiʃən] n. （RNA）翻译
[37] **splicing** ['splaisiŋ] n. 剪切，拼接

Exercises

1. Matching

1) subunit
2) deoxyribose
3) base pairing
4) transcription machinery
5) consensus sequence
6) enhancer
7) splicing

a) a modification of an RNA after transcription, in which introns are removed and exons are joined
b) a collection of a template, activated precursors, a divalent metalion, and RNA polymerase required by transcription
c) a short region of DNA that can increase transcription of genes
d) a pentose sugar $C_5H_{10}O_4$ that is a structural element of DNA
e) the rules specifying that guanine pairs with cytosine and adenine pairs with thymine
f) a subdivision of a larger unit
g) the most common nucleotide or amino acid at a particular position after multiple sequences are aligned

2. True or False

1) A nucleotide is composed of 2'-deoxyribose, a phosphate group, and one of the four bases adenine, cytosine, guanine, and thymine.
2) The base pairing between guanine and cytosine forms two hydrogen bonds.
3) One DNA polymer contains a 3' end and a 5' end.
4) RNA molecules are less stable than DNA and are typically single-stranded.
5) Many eukaryotic genes are organized into operons, or groups of genes whose products have related functions and which are transcribed as a unit.

3. Reading Comprehension

1) Which is not a base of RNA?
   A. adenine
   B. cytosine
   C. guanine
   D. thymine

2) Which of the following statements is not true?

A. The two strands in a DNA double helix must be complementary.
   B. Some genes have "strong" promoters and others have "weak" promoters.
   C. Prokaryotic genes may include long stretches of DNA called introns which are transcribed but never translated into protein.
   D. Splicing can also occur in prokaryotic genes, but is less common than in eukaryotes.
3) Which is a false description of a promoter?
   A. A gene can have no more than one promoter.
   B. Eukaryotic promoter regions are much more complex and difficult to identify than prokaryotic promoters.
   C. A promoter provides a position that is recognized by the transcription machinery.
   D. The weak promoters usually permit a lower rate of transcription than the strong promoters.

4. Translation from English to Chinese

In this structure, the base pairing rules specify that guanine pairs with cytosine and adenine pairs with thymine. The base pairing between guanine and cytosine forms three hydrogen bonds, whereas the base pairing between adenine and thymine forms two hydrogen bonds. The two strands in a double helix must therefore be complementary, that is, their bases must align such that the adenines of one strand are paired with the thymines of the other strand, and so on.

5. Translation from Chinese to English

基因可以有不止一个启动子，这就导致能从 5'端产生不同延伸长度的 RNA。启动子区有一个共有序列，在此位置序列保守性最高。有些基因具有与转录复合物结合非常好的强启动子，而另一些基因则只具有与转录复合物结合较差的弱启动子。因为转录复合物与这些弱启动子结合并起始转录的频率要低得多，所以它们的转录速度通常低于强启动子。其他的调控区可能还包括能弥补弱启动子的增强子。

6.3 Gene Expression

In all organisms, there are two major steps separating a protein-coding gene from its protein: First, the DNA on which the gene resides must be transcribed from DNA to messenger RNA (mRNA); and, second, it must be translated from mRNA to protein. RNA-coding genes must still go through the first step, but are not translated into protein. The process of producing a biologically functional molecule of either RNA or protein is called gene expression, and the resulting molecule itself is called a gene product[1].

Genetic code

The genetic code is the set of rules by which a gene is translated into a functional protein. Each gene consists of a specific sequence of nucleotides encoded in a DNA (or sometimes RNA) strand; a correspondence between nucleotides, the basic building blocks of genetic material, and amino acids, the basic building blocks of proteins, must be established for genes to be successfully translated into functional proteins. Sets of three nucleotides, known as codons, each correspond to a specific amino acid or to a signal; three codons are known as "stop codons[2]" and, instead of specifying a new amino acid, alert the translation machinery[3] that the end of the gene has been reached. There are 64 possible codons (four possible nucleotides at each of three positions, hence 61 possible codons) and only 20 standard amino acids; hence the code is redundant and multiple codons can specify the same amino acid. The correspondence between codons and amino acids is nearly universal among all known living organisms.

Transcription

The process of genetic transcription produces a single-stranded RNA molecule known as messenger RNA, whose nucleotide sequence is complementary to the DNA from which it was transcribed. The DNA strand whose sequence matches that of the RNA is known as the coding strand[4] and the strand from which the RNA was synthesized is the template strand[5]. Transcription is performed by an enzyme called an RNA polymerase[6], which reads the template strand in the 3' to 5' direction and synthesizes the RNA from 5' to 3'. To initiate transcription, the polymerase first recognizes and binds a promoter region of the gene. Thus a major mechanism of gene regulation[7] is the blocking or sequestering of the promoter region, either by tight binding by repressor[8] molecules that physically block the polymerase, or by organizing the DNA so that the promoter region is not accessible.

In prokaryotes, transcription occurs in the cytoplasm[9]; for very long transcripts[10], translation may begin at the 5' end of the RNA while the 3' end is still being transcribed. In eukaryotes, transcription necessarily occurs in the nucleus, where the cell's DNA is sequestered; the RNA molecule produced by the polymerase is known as the primary transcript[11] and must undergo post-transcriptional[12] modifications before being exported to the cytoplasm for translation. The splicing of introns present within the transcribed region is a modification unique to eukaryotes; alternative splicing mechanisms can result in mature transcripts[13] from the same gene having different sequences and thus coding for different proteins. This is a major form of regulation in eukaryotic cells.

Translation

Translation is the process by which a mature mRNA molecule is used as a template for synthesizing a new protein. Translation is carried out by ribosomes, large complexes of RNA and protein responsible for carrying out the chemical reactions to add new amino acids to a growing polypeptide chain[14] by the formation of peptide bonds[15]. The genetic code is read three nucleotides at a time, in units called codons, via interactions with specialized RNA molecules called transfer RNA (tRNA). Each tRNA has three unpaired bases known as the anticodon that are complementary to the codon it reads; the tRNA is also covalently attached to the amino acid specified by the complementary codon. When the tRNA binds to its complementary codon in an mRNA strand, the ribosome ligates its amino acid cargo to the new polypeptide chain, which is synthesized from amino terminus[16] to carboxyl terminus[17]. During and after its synthesis, the new protein must fold to its active three-dimensional structure before it can carry out its cellular function.

Notes to the Difficult Sentences

In eukaryotes, transcription occurs in the nucleus, where the cell's DNA is sequestered; the RNA molecule produced by the polymerase is known as the primary transcript and must undergo post-transcriptional modifications before being exported to the cytoplasm for translation. The splicing of introns present within the transcribed region is a modification unique to eukaryotes; alternative splicing mechanisms can result in mature transcripts from the same gene having different sequences and thus coding for different proteins. This is a major form of regulation in eukaryotic cells.

在真核生物中，转录发生在封闭 DNA 的细胞核中。聚合酶产生的 RNA 分子称作初级转录本，其在被运输到细胞质进行翻译之前必须经过转录后修饰。存在于转录区域的内含子的剪切是真核生物独有的一种修饰方式，改变剪切机制会使同一基因产生序列不同并且编码不同蛋白质的成熟转录本。这是存在于真核细胞的主要调控方式。

Professional Words and Phrases

[1]　**gene product**　基因产物
[2]　**stop codon**　终止密码子
[3]　**translation machinery**　翻译机
[4]　**coding strand**　编码链
[5]　**template strand**　模板链
[6]　**RNA polymerase**　RNA 聚合酶
[7]　**gene regulation**　基因调控
[8]　**repressor**　[ri'presə]　n. 阻遏物
[9]　**cytoplasm**　['saitəplæzəm]　n. 细胞质
[10]　**transcript**　['trænskript]　n. 转录本，转录物
[11]　**primary transcript**　初级转录本
[12]　**post-transcriptional**　[ˈpəust trænˈskripʃənəl]　adj. 转录后的
[13]　**mature transcript**　成熟转录本
[14]　**polypeptide chain**　多肽链
[15]　**peptide bond**　肽键
[16]　**amino terminus**　氨基端
[17]　**carboxyl terminus**　羧基端

Exercises

1. Matching

1) gene expression　　　a) DNA strand whose sequence matches mRNA
2) transcription　　　　b) an enzyme carrying out the transcription
3) stop codon　　　　　c) the process of DNA converting into mRNA
4) RNA polymerase　　d) a nucleotide triplet within messenger RNA that signals a termination of translation
5) coding strand　　　　e) the process of producing a biologically functional molecule of either

RNA or protein

2. True or False

1) RNA-coding genes can only be transcribed, but not translated.

2) The sequence of the template strand is almost the same with that of mRNA, but the thymines in place of uracils.

3) Transcription is the process by which a mature mRNA molecule is used as a template for synthesizing a new protein.

3. Reading Comprehension

1) What's the name of the transcript?

   A. message RNA

   B. transfer RNA

   C. ribosomal RNA

   D. none of the above

2) Which of the following statements is not true?

   A. Either RNA or protein resulting from gene expression is called a gene product.

   B. In prokaryotes, transcription occurs in the cytoplasm.

   C. There are 64 possible codons coding for only 20 standard amino acids.

   D. The splicing of introns present within the transcribed region is a modification unique to eukaryotes.

3) Which is not related to the transcription of a gene?

   A. the template strand

   B. RNA polymerase

   C. transfer RNA

   D. a promoter

4. Translation from English to Chinese

Transcription is performed by an enzyme called an RNA polymerase, which reads the template strand in the 3' to 5' direction and synthesizes the RNA from 5' to 3'. To initiate transcription, the polymerase first recognizes and binds a promoter region of the gene. Thus a major mechanism of gene regulation is the blocking or sequestering of the promoter region, either by tight binding by repressor molecules that physically block the polymerase, or by organizing the DNA so that the promoter region is not accessible.

5. Translation from Chinese to English

在所有生物中，遗传信息从编码蛋白的基因传递到蛋白质分为两步：第一步，基因所依赖的 DNA 必须从 DNA 被转录成信使 RNA；第二步，接着必须从 mRNA 被翻译成蛋白质。编码 RNA 的基因也必须经过第一步，但却不被翻译成蛋白质。产生具有生物学功能的 RNA 或蛋白质的过程称为基因表达，而所产生的分子则被称为基因产物。

6.4 Essentials for Genetic Engineering—Additional Reading

A clone is an exact copy of an organism, organ, single cell, organelle or macromolecule.

Cell lines for medical research are derived from a single cell allowed to replicate millions of times, producing masses of identical clones.

Gene cloning[1] is the act of making copies of a single gene. Once a gene is identified, clones can be used in many areas of biomedical and industrial research. Genetic engineering is the process of cloning genes into new organisms, or altering the DNA sequence to change the protein product. Genetic engineering depends on our ability to perform the following essential procedures.

1) Polymerase Chain Reaction[2]

The discovery of thermo-stable DNA polymerases[3], such as *Taq* polymerase, made it possible to manipulate DNA replication in the laboratory and was essential to the development of PCR. Primers[4] specific to a particular region of DNA, on either side of the gene of interest, are used, and replication is stopped and started repetitively, generating millions of copies of that gene. These copies can then be separated and purified using gel electrophoresis[5].

2) Restriction Enzymes[6]

The discovery of enzymes known as restriction endonucleases[7] has been essential to protein engineering. These enzymes cut DNA at specific locations based on the nucleotide sequence. Hundreds of different restriction enzymes, capable of cutting DNA at a distinct site, have been isolated from many different strains of bacteria. DNA cut with a restriction enzyme produces many smaller fragments, of varying sizes. These can be separated using gel electrophoresis or chromatography[8].

3) Visualizing DNA by Agarose Gel[9]

Purifying DNA from a cell culture, or cutting it using restriction enzymes wouldn't be of much use if we couldn't visualize the DNA—that is, find a way to view whether or not your extract contains anything, or what size fragments you've cut it into. One way to do this is by gel electrophoresis. Gels are used for a variety of purposes, from viewing cut DNA to detecting DNA inserts[10] and knockouts[11].

4) Join Two Pieces of DNA

In genetic research it is often necessary to link two or more individual strands of DNA, to create a recombinant strand, or close a circular strand that has been cut with restriction enzymes. Enzymes called DNA ligases[12] can create covalent bonds between nucleotide chains. The enzymes DNA polymerase I and polynucleotide kinase[13] are also important in this process, for filling in gaps, or phosphorylating[14] the 5' ends, respectively.

5) Selection of Small Self-Replicating DNA

Small circular pieces of DNA that are not part of a bacterial genome, but are capable of self-replication, are known as plasmids. Plasmids are often used as vectors[15] to transport genes between microorganisms. In biotechnology, once the gene of interest has been amplified and both the gene and plasmid are cut by restriction enzymes, they are ligated together generating what is known as a recombinant DNA. Viral (bacteriophage[16]) DNA can also be used as a vector, as can cosmids[17], recombinant plasmids containing bacteriophage genes.

6) Method to Move a Vector into a Host Cell

The process of transferring genetic material on a vector such as a plasmid, into new host cells, is called transformation. This technique requires that the host cells are exposed to an environmental change which makes them "competent[18]" or temporarily permeable to the vector. Electroporation[19] is one such technique. The larger the plasmid, the lower the efficiency with which it is taken up by cells. Larger DNA segments are more easily cloned using bacteriophage, retrovirus or other viral vectors or cosmids in a method called transduction. Phage or viral vectors are often used in regenerative medicine but may cause insertion of DNA in parts of our chromosomes where we don't want it, causing complications and even cancer.

7) Methods to Select Transgenic Organisms

Not all cells will take up DNA during transformation. It is essential that there be a method of detecting the ones that do. Generally, plasmids carry genes for antibiotic resistance and transgenic cells can be selected based on expression of those genes and their ability to grow on media containing that antibiotic. Alternative methods of selection depend on the presence of other reporter proteins[20] such as the x-gal/ lacZ system, or green fluorescence protein[21], which allow selection based on color and fluorescence, respectively.

Notes to the Difficult Sentences

In genetic research it is often necessary to link two or more individual strands of DNA, to create a recombinant strand, or close a circular strand that has been cut with restriction enzymes. Enzymes called DNA ligases can create covalent bonds between nucleotide chains. The enzymes DNA polymerase I and polynucleotide kinase are also important in this process, for filling in gaps, or phosphorylating the 5' ends, respectively.

在遗传学研究领域中，人们经常需要连接两个或多个单独的 DNA 链用于产生一个重组 DNA 链或是闭合一个已经用限制性酶切割的环形链。被称为 DNA 连接酶的酶能在核苷酸链之间生成共价键。DNA 聚合酶 I 和多核苷酸激酶在这一过程中也是非常重要的，分别被用于填补缺口或使 5'末端磷酸化。

Small circular pieces of DNA that are not part of a bacterial genome, but are capable of self-replication, are known as plasmids. Plasmids are often used as vectors to transport genes between microorganisms. In biotechnology, once the gene of interest has been amplified and both the gene and plasmid are cut by restriction enzymes, they are ligated together generating what is known as a recombinant DNA. Viral (bacteriophage) DNA can also be used as a vector, as can cosmids, recombinant plasmids containing bacteriophage genes.

一种非常小的环状 DNA 不是基因组的一部分，但能自主复制，被称作质粒。质粒经常被用作在微生物之间进行转运基因的载体。在生物技术领域，一旦令人感兴趣的基因被扩增并且基因和质粒被限制性酶切开，它们就被连接在一起进而产生重组 DNA。病毒（噬菌体）的 DNA 也能被用作载体，被称作黏粒，即含有噬菌体基因的质粒。

Professional Words and Phrases

[1] **gene cloning**　基因克隆
[2] **polymerase chain reaction**　聚合酶链式反应
[3] **DNA polymerase**　DNA 聚合酶
[4] **primer**　['praimə]　n. 引物
[5] **gel electrophoresis**　凝胶电泳

[6] **restriction enzyme** 限制性酶
[7] **restriction endonuclease** 限制性内切酶
[8] **chromatography** [krəumə'tɔgrəfi] n. 色谱法
[9] **agarose gel** 琼脂糖凝胶
[10] **DNA insert** DNA 插入
[11] **knockout** ['nɔkaut] n.（基因）敲除
[12] **DNA ligase** DNA 连接酶
[13] **polynucleotide kinase** 多核苷酸激酶
[14] **phosphorylate** ['fɔsfərileit] vt. 使磷酸化
[15] **vector** ['vektə] n. 载体
[16] **bacteriophage** [bæk'tiəriəʃəidʒ] n. 噬菌体
[17] **cosmid** ['kɔzmid] n. 黏粒
[18] **competent** ['kɔmpətənt] adj. 感受态的
[19] **electroporation** [i'lektrəupɔreiʃən] n. 电转化
[20] **reporter protein** 报告蛋白
[21] **green fluorescence protein** 绿色荧光蛋白

Exercises

1. Matching

1) gene cloning
2) restriction endonuclease
3) DNA ligase
4) bacteriophage
5) electroporation
6) cosmid
7) gel electrophoresis

a) any one of a number of viruses that infect bacteria
b) using a gel as an anticonvective medium and or sieving medium during electrophoresis
c) a significant increase in the electrical conductivity and permeability of the cell plasma membrane caused by an externally applied electrical field
d) a type of hybrid plasmid (often used as a cloning vector) that contains cos sequences
e) the act of making copies of a single gene
f) enzymes that cleave DNA at specific nucleotide sequences
g) a specific type of enzyme, a ligase, that repairs single-stranded discontinuities in double stranded DNA molecules

2. True or False

1) A clone is an exact copy of an organism, organ, single cell, organelle or macro-molecule.
2) All cells will take up DNA during transformation.
3) *Taq* polymerase is thermo-stable DNA polymerases.
4) The larger the plasmid, the higher the efficiency with which it is taken up by cells.
5) Restriction endonucleases often cut DNA at any locations based on the nucleotide sequence.

3. Reading Comprehension

1) Which can be used for a vector?
   A. plasmid
   B. cosmid

C. bacteriophage
   D. all

2) Which refers to the enzyme used for DNA ligation?
   A. restriction endonuclease
   B. DNA ligase
   C. DNA polymerase
   D. *Taq* polymerase

3) Which can be used for the selection of transgenic cells?
   A. antibiotic resistance
   B. x-gal/ lacZ system
   C. green fluorescence protein
   D. all

4. Translation from English to Chinese

Electroporation is one such technique. The larger the plasmid, the lower the efficiency with which it is taken up by cells. Larger DNA segments are more easily cloned using bacteriophage, retrovirus or other viral vectors or cosmids in a method called transduction. Phage or viral vectors are often used in regenerative medicine but may cause insertion of DNA in parts of our chromosomes where we don't want it, causing complications and even cancer.

5. Translation from Chinese to English

基因克隆指的是制造单基因拷贝的行为。一旦一个基因被鉴定，基因克隆就会被用在生物化学和工业研究中的诸多方面。基因工程是在新生物中对基因进行基因克隆的过程，或是通过改变DNA序列进而改变蛋白质产物的过程。基因工程取决于我们进行以下基本步骤的能力。

Chapter 7　Biochemistry

[本章中文导读]
　　生物化学是研究生命物质的化学组成、结构及生命过程中各种化学变化的科学，是生物学的重要分支学科之一。本章内容主要介绍了酶（第7.1节）、新陈代谢（第7.2节）、能量转化（第7.3节）以及蛋白质结晶（第7.4节）等生物化学方面重要的知识点。作者想通过利用英文讲解以上生物化学方面的专业基础知识以及生物化学领域的热点研究技术——蛋白质结晶等相关知识，达到使学生初步掌握生物化学的专业英语词汇的具体含义以及使用方法的目的。

7.1　Enzymes

　　Enzymes allow many chemical reactions to occur within the homeostasis constraints of a living system. Enzymes function as organic catalysts[1]. A catalyst is a chemical involved in, but not changed by, a chemical reaction. Many enzymes function by lowering the activation energy[2] of reactions. By bringing the reactants[3] closer together, chemical bonds[4] may be weakened and reactions will proceed faster than without the catalyst.

　　Enzymes can act rapidly, as in the case of carbonic[5] anhydrase[6] (enzymes typically end in the -ase suffix), which causes the chemicals to react 10^7 times faster than without the enzyme present. Carbonic anhydrase speeds up the transfer of carbon dioxide from cells to the blood. There are over 2000 known enzymes, each of which is involved with one specific chemical reaction. Enzymes are substrate[7] specific. The enzyme peptidase[8] (which breaks peptide bonds in proteins) will not work on starch (which is broken down by human-produced amylase[9] in the mouth).

　　Enzymes are proteins. The functioning of the enzyme is determined by the shape of the protein. The arrangement of molecules on the enzyme produces an area known as the active site[10] within which the specific substrate(s) will "fit". It recognizes, confines and orients the substrate in a particular direction.

　　The induced fit hypothesis[11] suggests that the binding of the substrate to the enzyme alters the structure of the enzyme, placing some strain on the substrate and further facilitating the reaction. Cofactors[12] are nonproteins essential for enzyme activity. Ions such as K^+ and Ca^{2+} are cofactors. Coenzymes[13] are nonprotein organic molecules bound to enzymes near the active site.

　　Enzymatic pathways form as a result of the common occurrence of a series of dependent chemical reactions. In one example, the end product depends on the successful completion of five reactions, each mediated by a specific enzyme. The enzymes in a series can be located adjacent to each other (in an organelle or in the membrane of an organelle), thus speeding the reaction process. Also, intermediate[14] products tend not to accumulate, making the process more efficient. By removing intermediates (and by inference end products[15]) from the reactive pathway, equilibrium (the tendency of reactions to reverse when concentrations of the products build up to a certain level) effects are minimized, since equilibrium is not attained, and so the reactions will proceed in the "preferred" direction.

Notes to the Difficult Sentences

Enzymes allow many chemical reactions to occur within the homeostasis constraints of a living system. Enzymes function as organic catalysts. A catalyst is a chemical involved in, but not changed by, a chemical reaction. Many enzymes function by lowering the activation energy of reactions. By bringing the reactants closer together, chemical bonds may be weakened and reactions will proceed faster than without the catalyst.

酶能允许许多化学反应发生在一个生命系统的体内平衡限制体系之内。酶是作为有机催化剂起作用的。催化剂是与化学反应有关但不改变化学反应的一种化学物质。许多酶是通过降低反应的活化能起作用的。通过使反应物相互接近到几乎在一起，化学键可以被削弱，而且反应本身将比没有催化剂存在的条件下进行得更快。

Professional Words and Phrases

[1] **catalyst** ['kætəlist] n. 催化剂
[2] **activation energy** 活化能
[3] **reactant** [ri'æktənt] n. 反应物
[4] **chemical bond** 化学键
[5] **carbonic** [kɑː'bɔnik] adj. 碳的
[6] **anhydrase** [æn'haidreis] n. 脱水酶
[7] **substrate** ['sʌbstreit] n. 底物
[8] **peptidase** ['peptideis] n. 肽酶
[9] **amylase** ['æməleis] n. 淀粉酶
[10] **active site** 活性部位
[11] **induced fit hypothesis** 诱导契合假说
[12] **cofactor** [kəu'fæktə] n. 辅因子
[13] **coenzyme** [kəu'enzaim] n. 辅酶
[14] **intermediate** [intə'miːdiət] n. 中间产物
[15] **end product** 终产物

Exercises

1. Matching

1) enzyme   a) the property of a system, either open or closed, that regulates its internal environment and tends to maintain a stable, constant condition

2) cofactor   b) a chemical compound synthesized from simpler compounds and usually intended to be used in later syntheses of more complex products

3) coenzyme   c) a substance that needs to be present in addition to an enzyme for a certain

　　　　　　　　　　　reaction to be catalysed
　　4) homeostasis　　d) a nonprotein organic molecule bound to enzymes near the active site
　　5) intermediate　　e) a type of organic catalyst

2. True or False

　　1) Carbonic anhydrase speeds up the transfer of carbon dioxide from cells to the blood.
　　2) The enzyme peptidase can be use to break open peptide bonds in the starch.
　　3) The intermediate products tend not to accumulate, making the process more efficient.

3. Reading Comprehension

　　1) Which is not the property of a enzyme?
　　　　A. Enzymes function as organic catalysts.
　　　　B. Enzymes are mainly proteins.
　　　　C. Enzymes are substrate non-specific.
　　　　D. Enzymes can act rapidly.
　　2) Which of the following statements is not true?
　　　　A. Ions such as K^+ and Ca^{2+} are cofactors.
　　　　B. There are over 2000 known enzymes.
　　　　C. A cofactor is the same to a coenzyme.
　　　　D. Many enzymes function by lowering the activation energy of reactions.
　　3) Which is the function of the enzyme amylase?
　　　　A. to speed up the transfer of carbon dioxide from cells to the blood.
　　　　B. to break peptide bonds in proteins.
　　　　C. to break down the starch.
　　　　D. to digest protein and remove contamination from preparations of nucleic acid.

4. Translation from English to Chinese

　　　The induced fit hypothesis suggests that the binding of the substrate to the enzyme alters the structure of the enzyme, placing some strain on the substrate and further facilitating the reaction. Cofactors are nonproteins essential for enzyme activity. Ions such as K^+ and Ca^{2+} are cofactors. Coenzymes are nonprotein organic molecules bound to enzymes near the active site.

5. Translation from Chinese to English

　　　酶是蛋白质。酶的功能是由酶的形状决定的。分子在酶中的排列可以产生一个称为活性部位的位置，专一性底物将结合在此。该位置可以识别、限制并按特定方向定位底物。

7.2　Metabolism

　　Metabolism is a set of processes performed by the living beings that allow them to interchange matter and energy with their environment. The living beings use the metabolic processes to grow and organize themselves internally. This is one of the visible manifestations of life. Unlike the biotic[1] (living) systems, some abiotic[2] (inert, without life) systems grow externally; for example, a copper chloride[3] crystal[4] grows toward outside adding material in its periphery or forming nuclei of growth around itself. The living beings grow internally although they do not reproduce, that is, even if they do not produce descendants. Therefore, the metabolism is a set of processes used by the living beings for maintaining their structure and molecular organization in a quasi-stable state[5].

<u>The metabolism consists of two basic types of interdependent phases, the phase in which the energy is freed and the phase in which energy is captured. The phase that consists of the disintegration of complex organic compound[6] to release energy is known as catabolism[7], whereas the phase that consists of the arrangement of organic compounds from simpler compounds to store energy is called anabolism[8].</u>

An example of catabolism is that a cell intakes a molecule of glucose from the surroundings and disintegrates it to release energy (glycolysis[9]). Immediately after the energy has been released, it is captured and stored by a specialized molecule, the adenosine triphosphate[10] or ATP. When the energy that was stored by the ATP is used in the synthesis of more complex compounds (e.g. bonding several monosaccharides[11] to build molecules of disaccharides[12] or polysaccharides[13]) the process is anabolism.

Generally, the materials obtained during the catabolic phase will be used in the anabolic phase. Many molecules are used to form cellular structures[14], others are used to transmit messages or information (signal transduction[15]) and others are used like energy sources to perform other cellular functions[16].

The carbohydrates constitute the main source of energy for all living beings. Lipids are the second source of energy, which could be used when the primary source is exhausted or independently of the exhaustion of the primary source of energy. Proteins would be the ultimate organic compounds to which the cells would resort like an energy source because most proteins function to regulate processes and/or to form structures. When an organism is compelled to make use of its proteins, it could be suffering from undernourishment.

Catabolism means disintegration, whereas anabolism means reorganization. The catabolism implies release of energy, whereas anabolism implies capture of energy. Catabolism implies disorganization of matter, whereas anabolism implies a more complex reorganization of matter.

Notes to the Difficult Sentences

The metabolism consists of two basic types of interdependent phases, the phase in which the energy is freed and the phase in which energy is captured. The phase that consists of the disintegration of complex organic compound to release energy is known as catabolism, whereas the phase that consists of the arrangement of organic compounds from simpler compounds to store energy is called anabolism.

代谢是由相互依赖的两个基本阶段组成的，即能量释放阶段和能量捕捉阶段。有机化合物分解并释放能量的阶段被称作分解代谢，而由较简单化合物排列成复杂的有机化合物并储存能量的阶段被称作合成代谢。

Professional Words and Phrases

[1] **biotic** [baiˈɔtik] adj. 生命的，生物的

[2] **abiotic** [eibai'ɔtik] adj. 非生物的
[3] **copper chloride** 氯化铜
[4] **crystal** ['kristəl] n. 晶体
[5] **quasi-stable state** 准稳态
[6] **organic compound** 有机化合物
[7] **catabolism** [kə'tæbəlizəm] n. 分解代谢
[8] **anabolism** [ə'nabə.lizəm] n. 合成代谢
[9] **glycolysis** [glai'kɔləsis] n. 糖酵解
[10] **adenosine triphosphate** 腺苷三磷酸
[11] **monosaccharide** [mɔnə'sækəraid] n. 单糖
[12] **disaccharide** [dai'sækəraid] n. 二糖
[13] **polysaccharide** [pɔli'sækə.raid] n. 多糖
[14] **cellular structure** 细胞结构
[15] **signal transduction** 信号转导
[16] **cellular function** 细胞功能

Exercises

1. Matching

1) metabolism   a) the arrangement of organic compounds from simpler compounds to store energy
2) catabolism   b) the metabolic pathway that converts glucose into pyruvate
3) anabolism   c) the disintegration of complex organic compound to release energy
4) glycolysis   d) a multifunctional nucleotide used in cells as a coenzyme
5) adenosine triphosphate e) a set of processes used by the living beings for maintaining their structure and molecular organization in a quasi-stable state

2. True or False

1) An example of anabolism is that a cell intakes a molecule of glucose from the surroundings and disintegrates it to release energy (glycolysis).
2) The energy form catabolism is immediately captured and stored by a specialized molecule, the adenosine triphosphate or ATP.
3) The process of bonding several monosaccharides to build molecules of disaccharides or polysaccharides is a type of anabolism.

3. Reading Comprehension

1) Which is a false description of the metabolism?
   A. Metabolism is a set of processes performed by the living beings.
   B. The metabolism consists of catabolism and anabolism.
   C. Metabolism is one of the visible manifestations of life.
   D. None of the above.
2) Which is not included in the catabolism?
   A. the glycolysis
   B. the disintegration of complex organic compound

C. protein synthesis

D. the process of energy release

3) Which is not included in the anabolism?

A. the arrangement of organic compounds

B. the storation of energy in ATP

C. lipid synthesis

D. the disintegration of carbohydrates

4. Translation from English to Chinese

An example of catabolism is when a cell intakes a molecule of glucose from the surroundings and disintegrates it to release energy (glycolysis). Immediately after the energy has been released, it is captured and stored by a specialized molecule, the adenosine triphosphate or ATP. When the energy that was stored by the ATP is used in the synthesis of more complex compounds (e.g. bonding several monosaccharides to build molecules of disaccharides or polysaccharides) the process is anabolism.

5. Translation from Chinese to English

分解代谢的意思是降解，而合成代谢的意思是重新组成。分解代谢意味着能量的释放，而合成代谢意味着能量的捕捉。分解代谢意味着物质的分解，而合成代谢意味着物质更复杂的重新组合。

7.3　Energy Transformation

Carbohydrate catabolism is the breakdown of carbohydrates into smaller units. Carbohydrates are usually taken into cells once they have been digested into monosaccharides. Once inside, the major route of breakdown is glycolysis, where sugars such as glucose and fructose[1] are converted into pyruvate[2] and some ATP is generated. Pyruvate is an intermediate in several metabolic pathways[3], but the majority is converted to acetyl-CoA[4] and fed into the citric acid cycle[5]. Although some more ATP is generated in the citric acid cycle, the most important product is NADH, which is made from NAD^+ as the acetyl-CoA is oxidized. This oxidation releases carbon dioxide as a waste product. In anaerobic conditions, glycolysis produces lactate[6], through the enzyme lactate dehydrogenase[7] re-oxidizing NADH to NAD^+ for re-use in glycolysis. An alternative route for glucose breakdown is the pentose phosphate pathway, which reduces the coenzyme NADPH and produces pentose sugars such as ribose, the sugar component of nucleic acids.

Fats are catabolised[8] by hydrolysis to free fatty acids[9] and glycerol[10]. The glycerol enters glycolysis and the fatty acids are broken down by beta oxidation[11] to release acetyl-CoA, which then is fed into the citric acid cycle. Fatty acids release more energy upon oxidation than carbohydrates because carbohydrates contain more oxygen in their structures.

Amino acids are either used to synthesize proteins and other biomolecules, or oxidized to urea[12] and carbon dioxide as a source of energy. The oxidation pathway starts with the removal of the amino group by a transaminase[13]. The amino group[14] is fed into the urea cycle[15], leaving a deaminated[16] carbon skeleton[17] in the form of a keto acid[18]. Several of these keto acids are intermediates in the citric acid cycle, for example the deamination[19] of glutamate[20] forms α-ketoglutarate[21]. The glucogenic[22] amino acids can also be converted into glucose, through gluconeogenesis[23].

In oxidative phosphorylation[24], the electrons removed from organic molecules are transferred to oxygen and the energy released is used to make ATP. This is done in eukaryotes by a series of proteins in the membranes of mitochondria called the electron transport chain[25]. In prokaryotes, these proteins are found in the cell's inner membrane. These proteins use the energy released from passing electrons from reduced molecules like NADH onto oxygen to pump protons across a membrane. Pumping protons out of the mitochondria creates a proton concentration difference across the membrane and generates an electrochemical gradient. This force drives protons back into the mitochondrion through the base of an enzyme called ATP synthase[26]. The flow of protons makes the stalk subunit rotate, causing the active site of the synthase[27] domain to change shape and phosphorylate adenosine diphosphate[28]—turning it into ATP.

Chemolithotrophy[29] is a type of metabolism found in prokaryotes where energy is obtained from the oxidation of inorganic compounds[30]. These organisms can use hydrogen, reduced sulfur compounds[31] (such as sulfide[32], hydrogen sulfide[33] and thiosulfate[34]), ferrous[35] iron (Fe II) or ammonia as sources of reducing power[36] and they gain energy from the oxidation of these compounds with electron acceptors[37] such as oxygen or nitrite[38]. These microbial processes are important in global biogeochemical cycles[39] such as acetogenesis[40], nitrification[41] and denitrification[42] and are critical for soil fertility.

In many organisms the capture of solar energy is similar in principle to oxidative phosphorylation, as it involves energy being stored as a proton concentration gradient and this proton motive force then driving ATP synthesis. The electrons needed to drive this electron transport chain come from light-gathering proteins called photosynthetic[43] reaction centres or rhodopsins[44]. Reaction centers are classed into two types depending on the type of photosynthetic pigment[45] present, with most photosynthetic bacteria only having one type, while plants and cyanobacteria have two.

In plants, algae, and cyanobacteria, photosystem II uses light energy to remove electrons from water, releasing oxygen as a waste product. The electrons then flow to the cytochrome[46] b6f complex, which uses their energy to pump protons across the thylakoid membrane in the chloroplast. These protons move back through the membrane as they drive the ATP synthase, as before. The electrons then flow through photosystem I and can then either be used to reduce the coenzyme $NADP^+$, for use in the Calvin cycle or recycled for further ATP generation.

Notes to the Difficult Sentences

Pyruvate is an intermediate in several metabolic pathways, but the majority is converted to acetyl-CoA and fed into the citric acid cycle. Although some more ATP is generated in the citric acid cycle, the most important product is NADH, which is made from NAD^+ as the acetyl-CoA is oxidized. This oxidation releases carbon dioxide as a waste product. In anaerobic conditions, glycolysis produces lactate, through the enzyme lactate dehydrogenase re-oxidizing NADH to NAD$^+$ for re-use in glycolysis. An alternative route for glucose breakdown is the pentose phosphate pathway, which reduces the coenzyme NADPH and produces pentose sugars such as ribose, the sugar component of nucleic acids.

丙酮酸是几个代谢途径的中间产物，但其大部分被转变成乙酰辅酶 A 而流入三羧酸循环。尽管三羧酸循环也产生一些 ATP，但是该循环最重要的产物是 NADH。该物质在乙酰辅酶 A 被氧化时由 NAD$^+$产生。该氧化作用释放废物二氧化碳。在厌氧条件下，糖酵解产生乳酸，乳酸可通过乳酸脱氢酶将 NADH 重新氧化成 NAD$^+$供糖酵解再次利用。葡萄糖降解的替代路线是磷酸戊糖途径，该途径还原辅酶 NADPH 并且产生核糖（核酸的糖成分）等戊糖。

Professional Words and Phrases

[1] **fructose** ['frʌktəus] n. 果糖
[2] **pyruvate** [pai'ru:veit] n. 丙酮酸盐
[3] **metabolic pathway** 代谢途径
[4] **acetyl-CoA** 乙酰辅酶 A
[5] **citric acid cycle** 柠檬酸循环，三羧酸循环
[6] **lactate** ['lækteit] n. 乳酸
[7] **lactate dehydrogenase** 乳酸脱氢酶
[8] **catabolise** [kə'tæbəlaiz] v.（使）分解代谢
[9] **fatty acid** 脂肪酸
[10] **glycerol** ['glisərəul] n. 甘油
[11] **beta oxidation** β-氧化作用
[12] **urea** [ju'ri:ə] n. 尿素
[13] **transaminase** [træn'sæmineis] n. 转氨酶
[14] **amino group** 氨基
[15] **urea cycle** 尿素循环
[16] **deaminate** [di'æməneit] v.（使）脱氨基
[17] **carbon skeleton** 碳骨架
[18] **keto acid** 酮酸
[19] **deamination** [di:æmi'neiʃən] n. 脱氨基作用
[20] **glutamate** ['glu:təmeit] n. 谷氨酸盐
[21] **α-ketoglutarate** ['ælfə ki:təu'glu:təreit] n. α-酮戊二酸
[22] **glucogenic** [glu:kə'dʒenik] adj. 生成葡糖的
[23] **gluconeogenesis** [glu:kəuni:ə'dʒenəsis] n. 糖异生作用
[24] **oxidative phosphorylation** 氧化磷酸化
[25] **electron transport chain** 电子传递链
[26] **ATP synthase** ATP 合成酶
[27] **synthase** ['sinθeis] n. 合成酶
[28] **adenosine diphosphate** 腺苷二磷酸
[29] **chemolithotrophy** [keməli'θɔtrəfi] n. 化能自养型

[30]　**inorganic compound**　无机化合物
[31]　**sulfur compound**　含硫化合物
[32]　**sulfide**　['sʌlfaid]　n. 硫化物
[33]　**hydrogen sulfide**　硫化氢
[34]　**thiosulfate**　[θaiəu'sʌlfeit]　n. 硫代硫酸盐
[35]　**ferrous**　['ferəs]　n. 亚铁，二价铁
[36]　**reducing power**　还原力
[37]　**electron acceptor**　电子受体
[38]　**nitrite**　['naitrait]　n. 亚硝酸盐
[39]　**biogeochemical cycle**　生物地球化学循环
[40]　**acetogenesis**　['æsitəu'dʒenəsis]　n. 乙酸形成作用
[41]　**nitrification**　[naitrəfi'keiʃən]　n. 硝化作用
[42]　**denitrification**　[di:naitrifi'keiʃən]　n. 脱氮作用，反硝化作用
[43]　**photosynthetic**　[fəutəusin'θetik]　adj. 光合的
[44]　**rhodopsin**　[rəu'dɔpsin]　n. 视紫质
[45]　**photosynthetic pigment**　光合色素
[46]　**cytochrome**　['saitəkrəum]　n. 细胞色素

Exercises

1. Matching

1) lactate dehydrogenase　　a) a metabolic pathway that uses energy released by the oxidation of nutrients to produce adenosine triphosphate

2) gluconeogenesis　　b) a process that generates NADPH and pentoses

3) oxidative phosphorylation　　c) an enzyme that catalyzes a type of reaction between an amino acid and an α-keto acid

4) transaminase　　d) an enzyme that catalyzes the conversion of lactate to pyruvate

5) pentose phosphate pathway　　e) a metabolic pathway that results in the generation of glucose from non-carbohydrate carbon substrates

2. True or False

1) The majority of pyruvate is converted to acetyl-CoA and fed into the citric acid cycle.
2) The oxidation pathway of amino acids starts with the removal of the amino group by a transaminase.
3) Chemolithotrophy is a type of metabolism found in eukaryotes where energy is obtained from the oxidation of inorganic compounds.

3. Reading Comprehension

1) Where is the ribose produced?

　　A. glycolysis

　　B. gluconeogenesis

　　C. pentose phosphate pathway

　　D. oxidative phosphorylation

2) Which way can ATP be produced?

A. oxidative phosphorylation

B. glycolysis

C. photosynthetic phosphorylation

D. all

3) Which of the following statements is not true?

A. The most photosynthetic bacteria only having one type of photosynthetic pigment, while plants and cyanobacteria have two.

B. ATP synthase can phosphorylate adenosine diphosphate to turn it into ATP.

C. In anaerobic conditions, more ATP is generated in the citric acid cycle.

D. Some prokaryotes can use hydrogen, reduced sulfur compounds ferrous iron (Fe II) or ammonia as sources of reducing power.

4. Translation from English to Chinese

Amino acids are either used to synthesize proteins and other biomolecules, or oxidized to urea and carbon dioxide as a source of energy. The oxidation pathway starts with the removal of the amino group by a transaminase. The amino group is fed into the urea cycle, leaving a deaminated carbon skeleton in the form of a keto acid. Several of these keto acids are intermediates in the citric acid cycle, for example the deamination of glutamate forms α-ketoglutarate. The glucogenic amino acids can also be converted into glucose, through gluconeogenesis.

5. Translation from Chinese to English

化能自养型是从氧化无机化合物获得能量的原核生物中发现的代谢类型。这些生物能利用氢还原含硫化合物（例如硫化物、硫化氢和硫代硫酸盐）、亚铁离子或氨等作为还原力并且利用氧或亚硝酸盐作为电子受体通过氧化这些化合物获得能量。这些微生物过程对乙酸形成、硝化作用和反硝化作用等全球生物地球化学循环非常重要，并且对土壤肥力非常关键。

7.4　Protein crystallization—Additional Reading

Proteins, like many molecules, can be prompted to form crystals when placed in the appropriate conditions. In order to crystallize a protein, the purified protein undergoes slow precipitation[1] from an aqueous solution[2]. As a result, individual protein molecules align themselves in a repeating series of unit cells by adopting a consistent orientation. The crystalline[3] lattice[4] that forms is held together by noncovalent interactions[5]. The importance of protein crystallization is that it serves as the basis for X-ray crystallography, wherein a crystallized protein is used to determine the protein's three-dimensional structure via X-ray diffraction[6].

The goal of crystallization is to produce a well-ordered crystal that is lacking in contaminants while still large enough to provide a diffraction pattern when exposed to X-rays. This diffraction pattern can then be analyzed to discern the protein's tertiary structure. Protein crystallization is inherently difficult because of the fragile nature of protein crystals[7]. Proteins have irregularly shaped surfaces, which results in the formation of large channels within any protein crystal. Therefore, the noncovalent bonds[8] that hold together the lattice must often be formed through several layers of solvent molecules.

In addition to overcoming the inherent fragility of protein crystals, a number of environmental factors must also be overcome. Due to the molecular variations between individual proteins, conditions unique to each protein must be obtained for a successful crystallization. Therefore, attempting to crystallize a protein without a proven protocol can be very challenging and time consuming.

Several conditions come into factors if a protein sample will crystallize or not. Some of these factors include protein purity, pH, protein concentration, temperature, and precipitants[9]. The more homogenous a protein in solution, the better the chances are for it to form a crystal. Typical standards have the protein solution being at least 97% pure. pH conditions are very important due to the fact that different pHs can result in different packing orientations. Buffers[10], such as Tris-HCl, are often necessary for the maintenance of a particular pH. Precipitants, such as ammonium sulfate[11] or polyethylene glycol[12], are compounds that cause the protein to precipitate out of solution.

<u>Two of the most commonly used methods for protein crystallization fall under the category of vapor diffusion[13]. These are known as the hanging drop method[14] and sitting drop methods[15]. Both entail a droplet containing purified protein, buffer, and precipitant being allowed to equilibrate with a larger reservoir containing similar buffers and precipitants in higher concentrations. Initially, the droplet of protein solution contains an insufficient concentration of precipitant for crystallization, but as water vaporizes from the drop and transfers to the reservoir, the precipitant concentration increases to a level optimal for crystallization.</u> Since the system is in equilibrium, these optimum conditions are maintained until the crystallization is complete. Simply put, the hanging drop method differs from the sitting drop method in the vertical orientation of the protein solution drop within the system. It is important to mention that both methods require a closed system, that is, the system must be sealed off from the outside using an airtight container or high-vacuum grease between glass surfaces.

Notes to the Difficult Sentences

Two of the most commonly used methods for protein crystallization fall under the category of vapor diffusion. These are known as the hanging drop method and sitting drop methods. Both entail a droplet containing purified protein, buffer, and precipitant being allowed to equilibrate with a larger reservoir containing similar buffers and precipitants in higher concentrations. Initially, the droplet of protein solution contains an insufficient concentration of precipitant for crystallization, but as water vaporizes from the drop and transfers to the reservoir, the precipitant concentration increases to a level optimal for crystallization.

最常用的两种蛋白质结晶方法都属于蒸汽扩散类型。它们是悬滴法和静滴法。两者都必须有一滴由纯化的蛋白质、缓冲液和沉淀剂组成的混合物，通过一个含相似缓冲液和较高浓度沉淀剂的大储水池进行平衡。开始时，蛋白质溶液液滴含有的沉淀剂浓度不足以使蛋白质结晶，但当水从蛋白质溶液液滴蒸发并转移到储水池中时，沉淀剂的浓度逐渐增加到最适浓度而导致结晶形成。

Professional Words and Phrases

[1] **precipitation** [prisipi'teiʃən] n. 沉淀
[2] **aqueous solution** 水溶液
[3] **crystalline** ['kristəlain] adj. 晶体的
[4] **crystalline lattice** 晶格
[5] **noncovalent interaction** 非供价作用
[6] **X-ray diffraction** X射线衍射
[7] **protein crystal** 蛋白质晶体
[8] **noncovalent bond** 非共价键
[9] **precipitant** [pri'sipitənt] n. 沉淀剂
[10] **buffer** ['bʌfə] n. 缓冲液
[11] **ammonium sulfate** 硫酸铵
[12] **polyethylene glycol** 聚乙二醇
[13] **vapor diffusion** 蒸汽扩散
[14] **hanging drop method** 悬滴法
[15] **sitting drop method** 静滴法

Exercises

1. Matching

1) protein crystallization
2) X-ray crystallography
3) precipitant
4) noncovalent bond
5) buffer

a) a substance or agent that causes a precipitate to form
b) a solution which reduces the change of pH upon addition of small amounts of acid or base, or upon dilution
c) a type of chemical bond that does not involve the sharing of pairs of electron
d) the method of making protein form a crystal
e) a method of determining the arrangement of atoms within a crystal, in which a beam of X-rays strikes a crystal and diffracts into many specific directions

2. True or False

1) The importance of protein crystallization is that it serves as the basis for X-ray crystallography.
2) Typical standards have the protein solution being at least 90% pure.
3) pH conditions are very important to protein crystallization due to the fact that different pHs can result in denature of proteins.
4) Due to the molecular variations between individual proteins, conditions unique to each protein must be obtained for a successful crystallization.
5) The more homogenous a protein in solution, the worse the chances are for it to form a crystal.

3. Reading Comprehension

1) Which is used for the determination of 3D structures of proteins?
   A. protein crystallization
   B. X-ray crystallography

C. vapor diffusion

D. noncovalent interaction

2) Which is not the factor affecting the protein crystallization?

A. protein purity

B. pH conditions

C. protein structure

D. temperature

3) Which of the following statements is not true?

A. Protein crystallization is inherently difficult because of the fragile nature of protein crystals.

B. Proteins can be prompted to form crystals when placed in the appropriate conditions.

C. The hanging drop and sitting drop methods for protein crystallization fall under the category of vapor diffusion.

D. The difference between hanging drop method and the sitting drop method is that the former does not need a closed system.

4. Translation from English to Chinese

The goal of crystallization is to produce a well-ordered crystal that is lacking in contaminants while still large enough to provide a diffraction pattern when exposed to X-rays. This diffraction pattern can then be analyzed to discern the protein's tertiary structure. Protein crystallization is inherently difficult because of the fragile nature of protein crystals. Proteins have irregularly shaped surfaces, which results in the formation of large channels within any protein crystal. Therefore, the noncovalent bonds that hold together the lattice must often be formed through several layers of solvent molecules.

5. Translation from Chinese to English

为了使蛋白质结晶，被纯化的蛋白质从水溶液中沉淀出来是非常慢的。结果是，单个的蛋白质分子按一致的方向自身排列成一系列的重复晶胞。形成的晶格是通过非共价作用结合在一起的。蛋白质结晶之所以重要是由于它是进行 X 射线衍射晶体学分析的基础，结晶的蛋白质借助 X 射线衍射晶体技术可以鉴定蛋白质的三维结构。

Chapter 8　Ecology

[本章中文导读]

　　生态学是研究生物之间以及生物与它们所在的周围环境之间关系的学科，是生物学重要分支学科之一。本章主要介绍了什么是生态学（第 8.1 节）、生态系统（第 8.2 节），以及生物多样性（第 8.3 节）等生态学的重要专业知识点。作者想通过利用英文讲解以上有关生态学的专业知识，达到使学生熟悉、掌握生态学中常见的专业英语词汇的具体含义和实际使用方法的目的。

8.1　What is Ecology?

　　Ecology is the scientific study of the relation of living organisms with each other and their surroundings. Ecosystems are defined by a web, community, or network of individuals that arrange into a self-organized and complex system and several levels of bigger systems constituting systems of smaller systems within them. Ecosystems create biophysical[1] feedback[2] between living (biotic) and nonliving (abiotic) components of an environment that generates and regulates the biogeochemical cycles[3] of the planet. Ecosystems provide goods and services that sustain human societies and general well-being. Ecosystems are sustained by biodiversity[4] within them. Biodiversity is the full-scale of life and its processes, including genes, species and ecosystems forming lineages that integrate into a complex and regenerative spatial arrangement of types, forms, and interactions.

　　Ecology is a sub-discipline of biology, the study of life. The word "ecology" was coined in 1866 by the German scientist Ernst Haeckel (1834—1919). Ancient philosophers of Greece, including Hippocrates and Aristotle, were among the earliest to record notes and observations on the natural history of plants and animals. Modern ecology later branched out of the natural history that flourished as a science in the late 19th century. Charles Darwin's evolutionary treatise including the concept of adaptation, as it was introduced in 1859, is a pivotal cornerstone in modern ecological theory.

　　Ecology is not synonymous with environment, environmentalism, natural history or environmental science[5]. It is closely related to physiology, evolutionary biology[6], genetics and ethology. An understanding of how biodiversity affects ecological function is an important focus area in ecological studies. Ecosystems sustain every life-supporting function on the planet, including climate regulation, water filtration[7], soil formation (pedogenesis[8]), food, fibers, medicines, erosion control, and many other natural features of scientific, historical or spiritual value.

　　Ecologists[9] seek to explain: ①life processes and adaptations, ②distribution and abundance of organisms, ③the movement of materials and energy through living communities, ④the successional development of ecosystems, ⑤the abundance and distribution of biodiversity in context of the environment.

There are many practical applications of ecology in conservation biology, wetland management, natural resource management (agriculture, forestry[10], fisheries), city planning (urban ecology[11]), community health, economics, basic & applied science and it provides a conceptual framework for understanding and researching human social interaction (human ecology[12]).

Notes to the Difficult Sentences

Ecosystems are defined by a web, community, or network of individuals that arrange into a self-organized and complex system and several levels of bigger systems constituting systems of smaller systems within them. Ecosystems create biophysical feedback between living (biotic) and nonliving (abiotic) components of an environment that generates and regulates the biogeochemical cycles of the planet. Ecosystems provide goods and services that sustain human societies and general well-being. Ecosystems are sustained by biodiversity within them.

生态系统的定义是在个体的网、群落或网络中，个体排列成的自身有组织性的复杂系统，以及具有数个层次的、其中含有小系统的更大系统。生态系统在环境的生物组分和非生物组分之间形成生物物理反馈，用以产生并调控我们这个星球的生物地球化学循环。生态系统可以提供维系人类社会和常规福利的食物和服务。生态系统则是通过其内部的生物多样性维系的。

Professional Words and Phrases

[1]　**biophysical**　[baiəuˈfizikəl]　adj. 生物物理的
[2]　**feedback**　[ˈfiːdbæk]　n. 反馈
[3]　**biogeochemical cycle**　生物地球化学循环
[4]　**biodiversity**　[baiəudaiˈvəːsəti]　n. 生物多样性
[5]　**environmental science**　环境科学
[6]　**evolutionary biology**　进化生物学
[7]　**filtration**　[filˈtreiʃən]　n. 过滤
[8]　**pedogenesis**　[piːdəuˈdʒenəsis]　n. 土壤发生
[9]　**ecologist**　[iːˈkɔlədʒist]　n. 生态学家
[10]　**forestry**　[ˈfɔristri]　n. 林业
[11]　**urban ecology**　城市生态学
[12]　**human ecology**　人类生态学

Exercises

1. Matching

1) biodiversity　　　　　a) a sub-field of biology concerned with the origin of species from a common descent and descent of species

2) pedogenesis　　　　　b) an interdisciplinary academic field that integrates physical and

	biological sciences to the study of the environment, and the solution of environmental problems
3) evolutionary biology	c) the full-scale of life and its processes
4) environmental science	d) the scientist who focus on the ecology
5) ecologist	e) soil formation

2. True or False

1) Modern ecology later branched out of the natural history that flourished as a science in the first 19th century.

2) Ecosystems are sustained by biodiversity within them.

3) Ecology is not synonymous with environment, environmentalism, natural history or environmental science.

3. Reading Comprehension

1) Who created the word "ecology"?
   A. Hippocrates
   B. Ernst Haeckel
   C. Aristotle
   D. Charles Darwin

2) Which is a true description of ecosystems?
   A. Ecosystems are a network of individuals that arrange into a self-organized and complex system.
   B. Ecosystems provide goods and services that sustain human societies and general well-being.
   C. Ecosystems sustain every life-supporting function on the earth.
   D. All.

3) What science was ecology ever included in?
   A. natural history
   B. environmental science
   C. physiology
   D. evolutionary biology

4. Translation from English to Chinese

Ecology is not synonymous with environment, environmentalism, natural history or environmental science. It is closely related to physiology, evolutionary biology, genetics and ethology. An understanding of how biodiversity affects ecological function is an important focus area in ecological studies. Ecosystems sustain every life-supporting function on the planet, including climate regulation, water filtration, soil formation (pedogenesis), food, fibers, medicines, erosion control, and many other natural features of scientific, historical or spiritual value.

5. Translation from Chinese to English

生态学家所要试图解释的是：①生命的过程及其适应性，②生物的分布及其丰度，③物质和能量在生命群落之间的流动，④生态系统的演替发展，⑤环境中生物多样性的丰度及其分布。

8.2 Ecosystems

Ecosystems are dynamic entities composed of the biological community and the abiotic environment. An ecosystem's abiotic and biotic composition and structure is determined by the state of a number of interrelated environmental factors. Changes in any of these factors (for example: nutrient availability, temperature, light intensity, grazing intensity, and species population density) will result in dynamic changes to the nature of these systems. For example, a fire in the temperate deciduous forest completely changes the structure of that system. There are no longer any large trees, most of the mosses, herbs, and shrubs that occupy the forest floor are gone, and the nutrients that were stored in the biomass are quickly released into the soil, atmosphere and hydrologic system[1]. After a short time of recovery, the community that was once large mature trees now becomes a community of grasses, herbaceous[2] species, and tree seedlings.

Ecosystems are confronted with natural environmental variations both over time and by differences between locations. The variations occur over vastly different ranges in terms of magnitudes as well as distances and time periods. It can take thousands of years for ecological processes to mature, for example, for all the successional stages of a forest. The area of an ecosystem can vary greatly from tiny to vast.

A single tree is of little consequence to the classification of a forest ecosystem, but critically relevant to the smaller organisms living in and on it. Several generations of an aphid[3] population can exist over the lifespan of a single leaf. Each of those aphids, in turn, support diverse bacterial communities. Fine scale structure of aphid populations can be constrained by influences from the growth of the tree, that is related to site specific variables, such as soil type, moisture content, slope of the land, and forest canopy closure, all phenomena on a much larger scale than the world of the aphid community. Likewise, finer scale dynamics operating in the aphid populations can impart influence on tree growth rates, i.e. influences from the small to the big. The scale of ecological dynamics[4] can operate as a closed island with respect to local site variables, such as aphids migrating on a tree, while at the same time remain open with regard to broader scale influences, such as atmosphere or climate. Hence, ecologists have devised means of hierarchically classifying ecosystems by analyzing data collected from finer scale units, such as vegetation associations[5], climate, and soil types, and integrate this information to identify larger emergent patterns of uniform organization and processes that operate on regional, local, and chronological scales.

There are different views on complexity and how it relates to ecology. One perspective lumps things that we do not understand into this category by virtue of the computational effort it would require to piece together the numerous interacting parts. Alternatively, complexity in life sciences can be viewed as emergent self-organized systems with multiple possible outcomes directed by random accidents of history; an extension of the first perspective. Global patterns of biological diversity are complex. This biocomplexity[6] stems from the interplay among ecological processes that operate and influence patterns that grade into each other, such as transitional areas or ecotones[7] that stretch across different scales. Complexity in ecology is of at least six distinct types: spatial, temporal, structural, process, behavioral, and geometric.

Small scale patterns do not necessarily explain large scale phenomena, otherwise captured in the expression "the sum is greater than the parts". Ecologists have identified emergent and self-organizing phenomena that operate at different environmental scales of influence, ranging from molecular to planetary[8], and these require different sets of scientific explanation. Long-term ecological studies provide important track records to better understand the complexity of ecosystems over longer temporal and broader spatial scales. The International Long Term Ecological Network manages and exchanges scientific information among research sites. The longest experiment in existence is the Park Grass Experiment that was initiated in 1856. Another example includes the Hubbard Brook study in operation since 1960.

To structure the study of ecology into a manageable framework of understanding, the biological world is conceptually organized as a nested hierarchy[9] of organization, ranging in scale from genes, to cells, to tissues, to organs, to organisms, to species and up to the level of the biosphere[10]. Together these hierarchical scales of life form a panarchy. Ecosystems are primarily researched at three key levels of organization—organisms, populations, and communities. Ecologists study ecosystems by sampling a certain number of individuals that are representative of a population. Ecosystems consist of communities interacting with each other and the environment. In ecology, communities are created by the interaction of the populations of different species in an area.

Notes to the Difficult Sentences

Ecosystems are dynamic entities composed of the biological community and the abiotic environment. An ecosystem's abiotic and biotic composition and structure is determined by the state of a number of interrelated environmental factors. Changes in any of these factors (for example: nutrient availability, temperature, light intensity, grazing intensity, and species population density) will result in dynamic changes to the nature of these systems.

生态系统是由生物群落和非生物环境因素组成的动态实体。生态系统的非生物组成和生物组成以及结构是由许多相互关联的环境因子的状态决定的。任何一个因子（例如，养分有效性、温度、光强度、放牧强度和物种种群密度）发生变化都将导致系统的本质发生动态改变。

Professional Words and Phrases

[1] **hydrologic system** 水文系统
[2] **herbaceous** [həːˈbeiʃəs] adj. 草本植物的
[3] **aphid** [ˈeifid] n. 蚜虫
[4] **ecological dynamics** 生态动力学
[5] **vegetation association** 植物群丛
[6] **biocomplexity** [ˈbaiəukəmˈpleksəti] n. 生物复杂性
[7] **ecotone** [ˈiːkətəun] n. 群落交错区
[8] **planetary** [ˈplænitəri] adj. 行星的

[9]　**nested hierarchy**　包含型等级系统
[10]　**biosphere**　[ˈbaiəsfiə]　n. 生物圈

Exercises

1. Matching

1) hydrologic system　　a) any of numerous very small soft-bodied homopterous insects that suck the juices of plants
2) aphid　　b) a transition area between two adjacent but different patches of landscape, such as forest and grassland
3) nested hierarchy　　c) the global sum of all ecosystems
4) ecotone　　d) a hierarchical ordering of nested sets
5) biosphere　　e) the entire cycle of water movement

2. True or False

1) An ecosystem's abiotic and biotic composition and structure is determined by the state of a number of interrelated environmental factors.
2) Ecosystems are primarily researched at three key levels of organization—organisms, populations, and communities.
3) A single tree is very critical to the classification of a forest ecosystem.

3. Reading Comprehension

1) Which is an environmental factor affecting the ecosystem's composition and structure?
   A. nutrient availability
   B. temperature
   C. species population density
   D. all

2) Which of the following statements is not true?
   A. The area of an ecosystem can vary greatly from tiny to vast.
   B. Changes in one or two environmental factors will not result in dynamic changes to the nature of these systems.
   C. Small scale patterns do not necessarily explain large scale phenomena.
   D. The communities are created by the interaction of the populations of different species in an area.

3) Which level is closest to the ecosystem?
   A. individuals
   B. populations
   C. communities
   D. none of the above

4. Translation from English to Chinese

　　The scale of ecological dynamics can operate as a closed island with respect to local site variables, such as aphids migrating on a tree, while at the same time remain open with regard to broader scale influences, such as atmosphere or climate. Hence, ecologists have devised means of hierarchically classifying ecosystems by analyzing data collected from finer scale units, such as vegetation associations, climate, and soil types, and integrate this information to identify larger emergent patterns of uniform organization and processes that operate on regional, local, and chronological scales.

5. Translation from Chinese to English

生态系统面临的是随时间和地点的不同而改变的自然环境变化。这些变化在量级、距离和时间段上都有很大的不同。生态过程的完善需要成千上万年，例如，一个森林的所有演替阶段。生态系统在面积上差异很大，面积从微小到巨大都可以。

8.3 Biodiversity—Additional Reading

The word "biodiversity" is a contracted version of "biological diversity". The Convention on Biological Diversity[1] defines biodiversity as: "the variability[2] among living organisms from all sources including, inter alia, terrestrial, marine and other aquatic ecosystems and the ecological complexes of which they are a part; this includes diversity within species, between species, and of ecosystems."

Thus, biodiversity includes genetic variation within species, the variety of species in an area, and the variety of habitat types within a landscape. Perhaps inevitably, such an all-encompassing definition, together with the strong emotive power of the concept, has led to somewhat cavalier use of the term biodiversity, in extreme cases to refer to life or biology itself. But biodiversity properly refers to the variety of living organisms.

Biological diversity is of fundamental importance to the functioning of all natural and human-engineered ecosystems, and by extension to the ecosystem services that nature provides free of charge to human society. Living organisms play central roles in the cycles of major elements (carbon, nitrogen, and so on) and water in the environment, and diversity is specifically important in that these cycles require numerous interacting species.

General interest in biodiversity has grown rapidly in recent decades, in parallel with the growing concern about nature conservation[3] generally, largely as a consequence of accelerating rates of natural habitat loss, habitat fragmentation[4] and degradation[5], and resulting extinctions of species. The IUCN Red List[6] estimates that 12%~52% of species within well-studied higher taxa[7] such as vertebrates and vascular plants are threatened with extinction. Based on data on recorded extinctions of known species over the past century, scientists estimate that current rates of species extinction are about 100 times higher than long-term average rates based on fossil data. Other plausible estimates suggest that present extinction rates now may have reached 1000 to 10,000 times the average over past geologic time. These estimates are the basis of the consensus that the Earth is in the midst of the sixth mass extinction event in its history; the present extinction event is termed the Holocene Mass Extinction.

Biodiversity is most frequently quantified as the number of species. Estimates of the number of species currently living on Earth range widely, largely because most living species are microorganisms and tiny invertebrates, but most estimates fall between 5 million and 30 million species. Roughly 1.75 million species have been formally described and given official names.

Insects comprise over half of the described species, and three fourths of known faunal species. The number of undescribed species is undoubtedly much higher, however. Particularly in inaccessible environments, and for inconspicuous groups of organisms, collecting expeditions routinely discover many undescribed species. Estimates of the total numbers of species on Earth have been derived variously by extrapolating from the ratios of described to previously unknown species in quantitative samples, from the judgment of experts in particular taxonomic groups, and from patterns in the description of new species through time. For most groups of organisms other than vertebrates, such estimates are a little more than educated guesses, explaining the wide range in estimates of global species diversity. Since insects are essentially absent from the sea, the species diversity of the oceans is generally considerably lower than terrestrial ones.

Species can be grouped on the basis of shared characteristics into hierarchical[8] groups, or taxa, reflecting their shared evolutionary history. At the highest level of classification (or deepest branches in the evolutionary tree[9] of life) organisms are divided into three Domains: ①the Bacteria[10], which are microorganisms lacking a cellular nucleus or other membrane-bound organelles; ② the relatively recently discovered Archaea[11], microorganisms of primarily extreme environments such as hot springs, which are superficially similar to Bacteria but fundamentally different at biochemical and genetic levels; and ③the Eukarya[12], which include all other organisms based on nucleated cells. The Eukarya includes the four "kingdoms[13]", the protists, animals, plants, and fungi. Each of the eukaryotic kingdoms[14] in turn is divided into a number of phyla[15]. At this higher taxonomic level, the oceans are far more diverse than those on land, likely reflecting the marine origins of life on Earth. Nearly half the phyla of animals occur only in the sea (e.g., the sea stars and other echinoderms[16]), whereas only one (the obscure Onychophora[17], or velvet worms[18]) is restricted to land.

Notes to the Difficult Sentences

At the highest level of classification (or deepest branches in the evolutionary tree of life) organisms are divided into three Domains: ①the Bacteria, which are microorganisms lacking a cellular nucleus or other membrane-bound organelles; ② the relatively recently discovered Archaea, microorganisms of primarily extreme environments such as hot springs, which are superficially similar to Bacteria but fundamentally different at biochemical and genetic levels; and ③the Eukarya, which include all other organisms based on nucleated cells.

在分类的最高水平（或生命进化树最底部的分支）上，生物被分成三域：①细菌域，即缺少细胞核和其他膜包被的细胞器的微生物；②最近才发现的古菌域，即主要是在像热泉等极端环境中存在的微生物，它们看上去类似细菌，但在生化和遗传水平上与细菌是完全不同的；③真核生物域，即以含有细胞核的细胞为基础的其他所有生命形式。

Professional Words and Phrases

[1] **Convention on Biological Diversity** 生物多样性公约
[2] **variability** [verɪəˈbɪləti] n. 变化性，变异性
[3] **nature conservation** 自然保护
[4] **habitat fragmentation** 生境破碎化

[5] **degradation** [degrəˈdeiʃən] n. 退化，降解
[6] **IUCN Red List** 世界自然保护联盟红色名录
[7] **taxa** [ˈtæksə] n. 分类单元，taxon 的复数形式
[8] **hierarchical** [ˌhaiəˈrɑːkikəl] adj. 按等级划分的
[9] **evolutionary tree** 进化树
[10] **Bacteria** [bækˈtiriə] n. 细菌域
[11] **Archaea** [ɑːˈkiə] n. 古菌域
[12] **Eukarya** [juːˈkæriə] n. 真核生物域
[13] **kingdom** [ˈkiŋdəm] n. （分类学上的）界
[14] **eukaryotic kingdom** 真核生物界
[15] **phyla** [ˈfailə] n. （分类学上的）门，phylum 的复数形式
[16] **echinoderm** [eˈkainədəːm] n. 棘皮动物类的动物
[17] **Onychophora** [ɔniˈkɔfərə] n. 有爪动物门
[18] **velvet worm** 栉蚕

Exercises

1. Matching

1) habitat fragmentation a) the world's most comprehensive inventory of the global conservation status of plant and animal species

2) IUCN Red List b) the emergence of discontinuities (fragmentation) in an organism's preferred environment

3) Bacteria c) organisms including all other organisms based on nucleated cells

4) Archaea d) a small phylum consisting of strange, caterpillar-like invertebrates that share traits with both arthropods and annelids

5) Eukarya e) microorganisms lacking a cellular nucleus or other membrane-bound organelles

6) Onychophora f) microorganisms of primarily extreme environments such as hot springs, which are superficially similar to Bacteria but fundamentally different at biochemical and genetic levels

2. True or False

1) Biodiversity properly refers to the variety of living organisms.

2) Insects comprise three fourths of the described species, and over half of known faunal species.

3) Archaea are superficially similar to Bacteria but fundamentally different at biochemical and genetic levels.

4) The IUCN Red List estimates that over 60% of species within vertebrates and vascular plants are threatened with extinction.

5) The species diversity of the oceans is generally similar to terrestrial ones.

3. Reading Comprehension

1) Which is included in the biodiversity?
   A. genetic variation within species

B. the variety of species in an area

   C. the variety of habitat types within a landscape

   D. all

2) Which is the number of species formally described and given official names?

   A. 5 million

   B. 30 million

   C. about 1.7 million

   D. 2 million

3) Which is not included in the domain Eukarya?

   A. animals

   B. bacteria

   C. plants

   D. fungi

4. Translation from English to Chinese

   The word "biodiversity" is a contracted version of "biological diversity". The Convention on Biological Diversity defines biodiversity as: "the variability among living organisms from all sources including, inter alia, terrestrial, marine and other aquatic ecosystems and the ecological complexes of which they are a part; this includes diversity within species, between species, and of ecosystems."

5. Translation from Chinese to English

   生物多样性对所有的自然生态系统和人为工程化的生态系统均是至关重要的，并且对自然界向人类社会无偿提供的生态系统服务功能也同样是至关重要的。生物对自然环境中的主要元素（碳、氮等）和水循环起到决定性的作用，生物多样性特别重要，因为这些循环需要各种各样的能够相互作用的物种。

Chapter 9　Biotechnology

[本章中文导读]

　　生物技术属于在工程学、医学等方面应用生命体及其生物过程的应用生物学领域。本章主要介绍了生物技术（第 9.1 节）、重组 DNA 技术（第 9.2 节）、重组蛋白表达技术（第 9.3 节）以及生物醇（第 9.4 节）等生物技术领域重要的专业基础知识，作者想通过利用英文讲解以上专业知识，达到使学生掌握生物技术领域重要的专业英语词汇的具体含义以及使用方法的目的。

9.1　Biotechnology Overview

　　Biotechnology[1] is a field of applied biology that involves the use of living organisms and bioprocesses[2] in engineering, technology, medicine and other fields requiring bioproducts[3]. Biotechnology also utilizes these products for manufacturing purpose. Modern use of similar terms includes genetic engineering as well as cell and tissue culture technologies. The concept encompasses a wide range of procedures (and history) for modifying living organisms according to human purposes — going back to domestication of animals, cultivation of plants, and "improvements" to these through breeding programs that employ artificial selection and hybridization[4]. By comparison to biotechnology, bioengineering[5] is generally thought of as a related field with its emphasis more on higher systems approaches (not necessarily altering or using biological materials directly) for interfacing with and utilizing living things. The United Nations Convention on Biological Diversity defines biotechnology as: "Any technological application that uses biological systems, living organisms, or derivatives[6], thereof, to make or modify products or processes for specific use." In other term "Application of scientific and technical advances in life science to develop commercial products" is biotechnology.

　　Biotechnology draws on the pure biological sciences (genetics, microbiology, animal cell culture, molecular biology, biochemistry, embryology[7], cell biology) and in many instances is also dependent on knowledge and methods from outside the sphere of biology (chemical engineering[8], bioprocess engineering[9], information technology[10], biorobotics[11]). Conversely, modern biological sciences (including even concepts such as molecular ecology[12]) are intimately entwined and dependent on the methods developed through biotechnology and what is commonly thought of as the life sciences[13] industry.

　　Biotechnology is not limited to medical/health applications (unlike Biomedical Engineering, which includes much biotechnology). Although not normally thought of as biotechnology, agriculture clearly fits the broad definition of "using a biotechnological[14] system to make products" such that the cultivation of plants may be viewed as the earliest biotechnological enterprise. Agriculture has been theorized to have become the dominant way of producing food since the Neolithic Revolution. The processes and methods of agriculture have been refined by other mechanical and biological sciences since its inception.

Through early biotechnology, farmers were able to select the best suited crops, having the highest yields, to produce enough food to support a growing population. Other uses of biotechnology were required as the crops and fields became increasingly large and difficult to maintain. Specific organisms and organism by-products were used to fertilize, restore nitrogen, and control pests. Throughout the use of agriculture, farmers have inadvertently altered the genetics of their crops through introducing them to new environments and breeding them with other plants—one of the first forms of biotechnology. Cultures such as those in Mesopotamia, Egypt, and India developed the process of brewing beer. It is still done by the same basic method of using malted grains (containing enzymes) to convert starch from grains into sugar and then adding specific yeasts to produce beer. In this process the carbohydrates in the grains were broken down into alcohols such as ethanol[15]. Ancient Indians also used the juices of the plant *Ephedra vulgaris*[16] and used to call it Soma. Later other cultures produced the process of lactic acid[17] fermentation[18] which allowed the fermentation and preservation of other forms of food. Fermentation was also used in this time period to produce leavened bread. Although the process of fermentation was not fully understood until Pasteur's work in 1857, it is still the first use of biotechnology to convert a food source into another form.

In the early twentieth century scientists gained a greater understanding of microbiology and explored ways of manufacturing specific products. In 1917, Chaim Weizmann first used a pure microbiological culture in an industrial process, that of manufacturing corn starch using *Clostridium acetobutylicum*[19], to produce acetone, which the United Kingdom desperately needed to manufacture explosives during World War I.

Biotechnology has also led to the development of antibiotics. In 1928, Alexander Fleming discovered the mold[20] *Penicillium*[21]. His work led to the purification of the antibiotic by Howard Florey, Ernst Boris Chain and Norman Heatley penicillin. In 1940, penicillin became available for medicinal use to treat bacterial infections in humans.

The field of modern biotechnology is thought to have largely begun on June 16, 1980, when the United States Supreme Court ruled that a genetically modified microorganism could be patented in the case of Diamond v. Chakrabarty. Indian-born Ananda Chakrabarty, working for General Electric, had developed a bacterium (derived from the *Pseudomonas*[22] genus[23]) capable of breaking down crude oil, which he proposed to use in treating oil spills.

Another factor influencing the biotechnology sector's success is improved intellectual property rights legislation—and enforcement—worldwide, as well as strengthened demand for medical and pharmaceutical[24] products to cope with an ageing, and ailing, U.S. population.

Rising demand for biofuels[25] is expected to be good news for the biotechnology sector, with the Department of Energy estimating ethanol usage could reduce U.S. petroleum-derived fuel consumption by up to 30% by 2030. The biotechnology sector has allowed the U.S. farming industry to rapidly increase its supply of corn and soybeans—the main inputs into biofuels—by developing genetically modified seeds which are resistant to pests and drought. By boosting farm productivity[26], biotechnology plays a crucial role in ensuring that biofuel production targets are met.

Notes to the Difficult Sentences

Rising demand for biofuels is expected to be good news for the biotechnology sector, with the Department of Energy estimating ethanol usage could reduce U.S. petroleum-derived fuel consumption by up to 30% by 2030. The biotechnology sector has allowed the U.S. farming industry to rapidly increase its supply of corn and soybeans—the main inputs into biofuels—by developing genetically modified seeds which are resistant to pests and drought. By boosting farm productivity, biotechnology plays a crucial role in ensuring that biofuel production targets are met.

可以预计人类对生物燃料需求的持续增长对于生物技术部来说是个好消息，因为美国能源部预计到 2030 年乙醇的使用将使美国石油衍生燃料的使用量减少 30%。生物技术部通过培育抗虫、抗旱种子，使美国农业增加对玉米和大豆的供应，这些作物主要用于生产生物燃料。随着农业生产力的提高，生物技术将在确保实现生物燃料生产目标达成方面起到关键作用。

Professional Words and Phrases

[1] **biotechnology** [baiəutek'nɔlədʒi] n. 生物技术
[2] **bioprocess** [baiəu'prɑːses] n. 生物处理，生物过程
[3] **bioproduct** [baiəu'prɔdəkt] n. 生物制品
[4] **hybridization** [haibridai'zeiʃən] n. 杂交
[5] **bioengineering** [baiəuendʒə'niəriŋ] n. 生物工程，生物工程学
[6] **derivative** [di'rivətiv] n. 衍生物
[7] **embryology** [embri'ɔlədʒi] n. 胚胎学
[8] **chemical engineering** 化学工程学
[9] **bioprocess engineering** 生物加工过程
[10] **information technology** 信息技术
[11] **biorobotics** [baiəurəu'bɔtiks] n. 仿生机器人学
[12] **molecular ecology** 分子生态学
[13] **life science** 生命科学
[14] **biotechnological** [baiəutek'nɔlədʒikəl] adj. 生物技术的
[15] **ethanol** ['eθənəul] n. 乙醇
[16] *Ephedra vulgaris* 麻黄
[17] **lactic acid** 乳酸
[18] **fermentation** [fəːmen'teiʃən] n. 发酵
[19] *Clostridium acetobutylicum* 丙酮丁醇梭菌
[20] **mold** [məuld] n. 霉菌
[21] *Penicillium* [peni'siliəm] n. 青霉菌属
[22] *Pseudomonas* [(p)sjuː'dɔmənəs] n. 假单胞菌属
[23] **genus** ['dʒiːnəs] n.（分类学上的）属
[24] **pharmaceutical** [fɑːmə'sjuːtikəl] adj. 制药的；n. 药品
[25] **biofuel** [baiəu'fjuːəl] n. 生物燃料

[26] farm productivity   农业生产力

Exercises

1. Matching

1) biotechnology   a) a science which is about the development of an embryo from the fertilization of the ovum to the fetus stage
2) hybridization   b) a group of antibiotics derived from mold *Penicillium*
3) embryology   c) a wide range of fuels which are in some way derived from biomass
4) fermentation   d) the process of combining different varieties or species of organisms to create a hybrid
5) penicillin   e) a metabolic process whereby electrons released from nutrients are ultimately transferred to molecules obtained from the breakdown of those same nutrients
6) biofuel   f) a field of applied biology that involves the use of living organisms and bioprocesses in engineering, technology, medicine and other fields requiring bioproducts

2. True or False

1) Biotechnology refers to the application of scientific and technical advances in life science to develop commercial products.
2) Alexander Fleming discovered the mold *Penicillium* and purified the antibiotic penicillin.
3) The process of fermentation was not fully understood until Pasteur's work in 1857.
4) In 1917, Chaim Weizmann first used a pure microbiological culture in an industrial process.
5) Bioengineering led to the development of antibiotics.

3. Reading Comprehension

1) Which may be viewed as the earliest biotechnological enterprise?
    A. medicine
    B. agriculture
    C. biochemical engineering
    D. genetic engineering
2) Which of the following statements is not true?
    A. Biotechnology is equal to bioengineering.
    B. Biotechnology draws on the pure biological sciences.
    C. Modern biological sciences are dependent on the methods developed through biotechnology.
    D. The field of modern biotechnology is thought to have largely begun on June 16, 1980.
3) Who had developed a bacterium capable of breaking down crude oil?
    A. Howard Florey
    B. Ananda Chakrabarty
    C. Chaim Weizmann
    D. Ernst Boris Chain

4. Translation from English to Chinese

Biotechnology draws on the pure biological sciences (genetics, microbiology, animal cell culture, molecular biology, biochemistry, embryology, cell biology) and in many instances is also dependent on knowledge and methods from outside the sphere of biology (chemical engineering, bioprocess engineering, information technology, biorobotics). Conversely, modern biological sciences (including even concepts such as molecular ecology) are intimately entwined and dependent on the methods developed through biotechnology and what is commonly thought of as the life sciences industry.

5. Translation from Chinese to English

生物技术不仅仅局限于在医学、健康学领域的应用（不像包括许多生物技术的生物医学工程）。尽管农业通常来说不被认为是生物技术，但是农业的确符合广义上的"使用生物技术体系生产产物"的定义，以至于植物的耕作可以视为生物技术的最早行业。

9.2 Recombinant DNA Technology

Recombinant DNA[1] (rDNA) molecules are nucleic acid sequences that result from the use of laboratory methods to bring together genetic material from multiple sources, creating sequences that would not otherwise be found in biological organisms. Recombinant DNA is made possible because DNA molecules from all organisms share the same chemical structure; they differ only in the sequence of nucleotides within that identical overall structure. Therefore, when DNA from a foreign source is hooked up to host sequences that can drive DNA replication[2] and then introduced into a host organism, the foreign DNA is replicated along with the host DNA.

The DNA sequences used in the construction of recombinant DNA molecules can originate from any species. For example, plant DNA may be joined to bacterial DNA, or human DNA may be joined with fungal DNA. In addition, DNA sequences that do not occur anywhere in nature may be created by the chemical synthesis of DNA, and incorporated into recombinant molecules. Using recombinant DNA technology and synthetic DNA[3], literally any DNA sequence may be created and introduced into any of a very wide range of living organisms.

It is important to note that recombinant DNA differs from genetic recombination[4] in that the former results from artificial methods in the test tube, while the latter is a normal biological process that results in the remixing of existing DNA sequences in essentially all organisms. Proteins that result from the expression of recombinant DNA within living cells are termed recombinant proteins[5].

Recombinant DNA technology begins with the isolation of a gene of interest. The gene is then inserted into a vector and cloned. A vector is a piece of DNA that is capable of independent growth; commonly used vectors are bacterial plasmids and viral phages[6]. The gene of interest (foreign DNA[7]) is integrated into the plasmid or phage, and this is referred to as recombinant DNA.

Before introducing a vector containing foreign DNA into host cells[8] to express a protein, it must be cloned. Cloning is necessary to produce numerous copies of the DNA since the initial supply is inadequate to insert into host cells.

Once the vector is isolated in large quantities, it can be introduced into the desired host cells, such as mammalian[9], yeast or special bacterial cells. The host cells will then synthesize the foreign protein from the recombinant DNA. When the cells are grown in vast quantities, the foreign or recombinant protein can be isolated and purified in large amounts.

The recombinant DNA technology was first proposed by Peter Lobban, a graduate student, with A. Dale Kaiser at the Stanford University Department of Biochemistry. Exploitation of recombinant DNA technology was facilitated by the discovery, isolation and application of restriction endonucleases by Werner Arber, Daniel Nathans, and Hamilton Smith, for which they received the 1978 Nobel Prize in Medicine. Cohen and Boyer applied for a patent on the process for producing biologically functional molecular chimeras[10] which could not exist in nature in 1974. A breakthrough in the application of recombinant DNA technology occurred in 1977 when Herbert Boyer produced biosynthetic[11] "human" insulin[12] in the lab. The specific gene sequence[13], or polynucleotide[14], that codes for insulin production in humans was introduced to a sample colony[15] of the *Escherichia coli*[16] bacteria. It was the first medicine made via recombinant DNA technology to be approved by the FDA and commercially available under the brand name Humulin. The vast majority of insulin currently used worldwide is now biosynthetic recombinant "human" insulin or its analogs[17].

Notes to the Difficult Sentences

Recombinant DNA is made possible because DNA molecules from all organisms share the same chemical structure; they differ only in the sequence of nucleotides within that identical overall structure. Therefore, when DNA from a foreign source is hooked up to host sequences that can drive DNA replication and then introduced into a host organism, the foreign DNA is replicated along with the host DNA.

重组 DNA 之所以有可能获得，是因为不同生物的 DNA 分子具有同样的化学结构；唯一不同的是在相同结构的前提下核苷酸在序列上的差异。因此，当外源 DNA 与能够完成 DNA 复制的宿主序列结合并被引入宿主生物中时，外源 DNA 可以随宿主 DNA 一起得以复制。

Professional Words and Phrases

[1] **recombinant DNA** 重组 DNA
[2] **DNA replication** DNA 复制
[3] **synthetic DNA** 合成 DNA
[4] **genetic recombination** 遗传重组
[5] **recombinant protein** 重组蛋白
[6] **phage** [feidʒ] n. 噬菌体
[7] **foreign DNA** 外源 DNA
[8] **host cell** 宿主细胞
[9] **mammalian** [mæ'meiljən] adj. 哺乳动物的
[10] **chimera** [kai'mirə] n. 嵌合体
[11] **biosynthetic** [baiəusin'θetik] adj. 生物合成的

[12] **insulin** [ˈinsjulin] n. 胰岛素
[13] **gene sequence** 基因序列
[14] **polynucleotide** [pɔliˈnjuːkljətaid] n. 多聚核苷酸
[15] **colony** [ˈkɔləni] n. 菌落
[16] ***Escherichia coli*** 大肠杆菌
[17] **analog** [ˈænəlɔːg] n. 类似物，同系物

Exercises

1. Matching

1) foreign DNA   a) a piece of circular DNA that is capable of independent growth
2) recombinant DNA   b) a biopolymer composed of 13 or more nucleotide monomers covalently bonded in a chain
3) recombinant protein   c) of, related to a mammal
4) vector   d) a hormone central to regulating carbohydrate and fat metabolism in the body
5) polynucleotide   e) a protein that result from the expression of recombinant DNA within living cells
6) mammalian   f) nucleic acid sequences to bring together genetic material from multiple sources
7) insulin   g) DNA from a foreign source

2. True or False

1) Restriction endonucleases were first discovered, isolated and used by Werner Arber, Daniel Nathans, and Hamilton Smith.
2) Cloning is necessary to produce numerous copies of the DNA.
3) DNA sequences can't be created by the chemical synthesis of DNA.

3. Reading Comprehension

1) Which is able to be used as a foreign DNA?
   A. plant DNA
   B. human DNA
   C. synthetic DNA
   D. all

2) Who first proposed the recombinant DNA technology?
   A. Werner Arber
   B. Peter Lobban
   C. Daniel Nathans
   D. Hamilton Smith

3) Which can be used as the host cells?
   A. mammalian cells
   B. yeast cells
   C. bacterial cells
   D. all

4. Translation from English to Chinese
Once the vector is isolated in large quantities, it can be introduced into the desired host cells, such as mammalian, yeast or special bacterial cells. The host cells will then synthesize the foreign protein from the recombinant DNA. When the cells are grown in vast quantities, the foreign or recombinant protein can be isolated and purified in large amounts.

5. Translation from Chinese to English
需要着重指出的是重组 DNA 技术不同于遗传重组，因为前者是在试管中采用人工方法完成的，而后者是导致所有生物中已知 DNA 序列进行重新组合的一个正常生物的过程。由活细胞中重组 DNA 表达产生的蛋白质被称为重组蛋白。

9.3 Recombinant Protein Expression

When you want to characterize[1] a gene or protein of interest, you must first study its function. In this molecular era, obtaining a cDNA[2] of your gene of interest is not difficult. To express the cDNA as a protein, ie. a recombinant protein, one can then easily perform functional studies using the recombinant purified protein.

There are two main systems for the expression of recombinant protein. Once you get your cDNA cloned, you must decide where you want to amplify[3] your protein. This will be either a prokaryotic[4] (bacterial) or eukaryotic (usually yeast or mammalian cell[5]) system. The choice of your system will decide which vector you will need to clone your cDNA into as there are different promoters which function in *Escherichia coli* and others that work best with yeast or mammalian systems.

Prokaryotic recombinant protein expression systems have several advantages. These include ease of culture, and very rapid cell growth meaning you won't have to wait long to get protein from bacterial systems once you clone your cDNA. Expression can be induced easily in bacterial protein expression systems using IPTG[6]. Also, purification is quite simple in prokaryotic expression systems and there are a plethora of commercial kits[7] available for recombinant protein expression.

On the other hand, if you need to use your proteins for functional or enzymatic studies, prokaryotic systems are a problem as most proteins become insoluble in inclusion bodies[8] and are very difficult to recover as functional proteins. Furthermore, most if not all post-translational modifications[9] are not added by bacteria and therefore your protein of interest may not be functional. Enzymatic studies thus may be unfruitful.

Eukaryotic genes are not really "at home" in prokaryotic cells, even when they are expressed under the control of the prokaryotic vectors. One reason is that *E. coli* cells frequently recognize the protein products of cloned eukaryotic genes as outsiders and destroy them. Another is that prokaryotes do not carry out the same kinds of post-translational modification as eukaryotes do. For example, a protein that would ordinarily be coupled to sugars in a eukaryotic cell will be expressed as a bare protein when cloned in bacteria. This can affect a protein's activity or stability, or at least its response to antibodies. A more serious problem is that the interior of a bacterial cell is not as conducive to proper folding of eukaryotic proteins as the interior of a eukaryotic cell. Frequently, the result is improperly folded, inactive products of cloned genes.

Eukaryotic systems for the expression of protein include: ①yeast, ②mammalian cells, ③baculovirus cells[10] (insect). All these systems are great eukaryotic systems for the expression of recombinant proteins.

Advantages of eukaryotic protein expression systems include the fact that you can get very high levels of expression. The proteins are easy to purify using special tags which are included into the vectors including Histag[11], Myctag[12] and other tags.

You can even purchase plasmids which secrete your protein into the media. Therefore you can keep growing your system and collecting the media without lysing[13] your cells. There are no inclusion bodies to worry about and your proteins have intact post-translational modifications. These are vital if you are studying the function of a protein and/or protein-protein interactions[14].

The disadvantages of eukaryotic protein expression systems include the fact that eukaryotic cells do grow slower than prokaryotic cells.

Notes to the Difficult Sentences

One reason is that *E. coli* cells frequently recognize the protein products of cloned eukaryotic genes as outsiders and destroy them. Another is that prokaryotes do not carry out the same kinds of post-translational modification as eukaryotes do. For example, a protein that would ordinarily be coupled to sugars in a eukaryotic cell will be expressed as a bare protein when cloned in bacteria. This can affect a protein's activity or stability, or at least its response to antibodies. A more serious problem is that the interior of a bacterial cell is not as conducive to proper folding of eukaryotic proteins as the interior of a eukaryotic cell.

原因之一是大肠杆菌细胞经常把所克隆的真核基因的蛋白质产物识别为外源物质并将它们降解。另一个原因是原核生物并不进行真核生物所具有的翻译后修饰。例如，当被克隆到细菌中时，通常在真核细胞中偶联有糖的蛋白质将仅表达蛋白质本身，而无糖的偶联。这会影响其蛋白质活性、稳定性或至少是抗体的响应能力。更严重的问题是细菌细胞内部不像真核细胞那样适合真核蛋白的正确折叠。

Professional Words and Phrases

[1] **characterize** ['kæriktəraiz] v. 鉴定
[2] **cDNA** 互补 DNA，complementary DNA 的缩写
[3] **amplify** ['æmplifai] vt. 扩增
[4] **prokaryotic** [prəukæri'ɔtik] adj. 原核的
[5] **mammalian cell** 哺乳动物细胞
[6] **IPTG** 异丙基硫代-β-D-半乳糖苷（isopropyl-beta-D-thiogalactopyranoside）
[7] **kit** [kit] n. 试剂盒
[8] **inclusion body** 包涵体
[9] **post-translational modification** 翻译后修饰
[10] **baculovirus cell** 杆状病毒细胞
[11] **His tag** 组氨酸标签

[12] **Myc tag**　Myc（一个癌基因）标签
[13] **lyse**　[laɪs]　v. 裂解（细胞）
[14] **protein-protein interaction**　蛋白质与蛋白质之间的相互作用

Exercises

1. Matching

1) His tags
2) inclusion bodies
3) IPTG

4) post-translational modification
5) purification

6) cDNA

a) the process of rendering something pure
b) the abbr. of isopropyl-beta-D-thiogalac-topyranoside
c) DNA synthesized from a mature mRNA template in a reaction catalyzed by the enzyme reverse transcriptase and the enzyme DNA polymerase
d) the chemical modification of a protein after its translation
e) nuclear or cytoplasmic aggregates of stainable substances, usually proteins
f) an amino acid motif in proteins that consists of at least five histidine (His) residues

2. True or False

1) In the eukaryotic systems, most proteins are easy to become insoluble in inclusion bodies.
2) *E. coli* cells frequently recognize the protein products of cloned eukaryotic genes as outsiders and destroy them.
3) The interior of a bacterial cell is not as conducive to proper folding of eukaryotic proteins as the interior of a eukaryotic cell.

3. Reading Comprehension

1) Which is frequently used for the prokaryotic systems?
   A. *E. coli*
   B. yeast
   C. mammalian cells
   D. none of the above

2) Which is used for the eukaryotic systems?
   A. yeast
   B. mammalian cells
   C. baculovirus cells
   D. all

3) Which of the following statements is not true?
   A. Eukaryotic genes are not often expressed properly in prokaryotic cells.
   B. Eukaryotic cells do grow faster than prokaryotic cells.
   C. Both prokaryotic systems and eukaryotic systems have some advantages and disadvantages.
   D. In prokaryotic systems, protein expression can be induced easily using IPTG.

4. Translation from English to Chinese

　　There are two main systems for the expression of recombinant protein. Once you get your cDNA cloned, you must decide where you want to amplify your protein. This will be either a prokaryotic (bacterial) or eukaryotic (usually yeast or mammalian cell) system. The choice of your system will decide which vector you will need to clone your cDNA into as there are different promoters which function in *Escherichia coli* and others that work best with yeast or mammalian systems.

5. Translation from Chinese to English

原核重组蛋白表达系统具有几个优点，包括细胞易培养并且生长快速，这意味着一旦克隆 cDNA，将不必等很长时间就可以从细菌表达系统获得蛋白质。使用 IPTG 很容易在细菌蛋白表达系统中诱导蛋白表达。此外，在原核表达系统中纯化也相当简单，并且有很多商用试剂盒可以用于重组蛋白的表达。

9.4 Bioalcohols—Additional Reading

Biologically produced alcohols (bioalcohol[1]), most commonly ethanol[2], and less commonly propanol[3] and butanol[4], are produced by the action of microorganisms and enzymes through the fermentation of sugars or starches (easiest), or cellulose[5] (which is more difficult). Biobutanol[6] (also called biogasoline[7]) is often claimed to provide a direct replacement for gasoline, because it can be used directly in a gasoline engine (in a similar way to biodiesel[8] in diesel engines).

Bioethanol[9] is an alcohol made by fermenting[10] the sugar components of plant materials and it is made mostly from sugar and starch crops. With advanced technology being developed, cellulosic[11] biomass, such as trees and grasses, are also used as feedstocks[12] for ethanol production. Ethanol can be used as a fuel for vehicles in its pure form, but it is usually used as a gasoline additive to increase octane and improve vehicle emissions. Bioethanol is widely used in the USA and in Brazil.

Ethanol fuel is the most common biofuel worldwide, particularly in Brazil. Alcohol fuels are produced by fermentation of sugars derived from wheat, corn, sugar beets[13], sugar cane[14], molasses[15] and any sugar or starch that alcoholic beverages can be made from (like potato and fruit waste, etc.). The ethanol production methods used are enzyme digestion (to release sugars from stored starches), fermentation of the sugars, distillation[16] and drying. The distillation process requires significant energy input for heat (often unsustainable natural gas fossil fuel, but cellulosic biomass such as bagasse[17], the waste left after sugar cane is pressed to extract its juice, can also be used more sustainably).

Ethanol can be used in petrol engines as a replacement for gasoline; it can be mixed with gasoline to any percentage. Most existing car petrol engines can run on blends of up to 15% bioethanol with petroleum/gasoline. Ethanol has a smaller energy density than gasoline, which means it takes more fuel (volume and mass) to produce the same amount of work. An advantage of ethanol (CH_3CH_2OH) is that it has a higher octane[18] rating than ethanol-free gasoline available at roadside gas stations which allows an increase of an engine's compression ratio for increased thermal efficiency[19]. In high altitude (thin air) locations, some states of the United States mandate a mix of gasoline and ethanol as a winter oxidizer[20] to reduce atmospheric pollution emissions.

Ethanol is also used to fuel bioethanol fireplaces. As they do not require a chimney and are "flueless", bioethanol fires are extremely useful for new build homes and apartments without a flue. The downside to these fireplaces, is that the heat output is slightly less than electric and gas fires.

In the current alcohol-from-corn production model in the United States, considering the total energy consumed by farm equipment, cultivation, planting, fertilizers, pesticides, herbicides, and fungicides[21] made from petroleum, irrigation systems, harvesting, transport of feedstock to processing plants, fermentation, distillation, drying, transport to fuel terminals and retail pumps, and lower ethanol fuel energy content, the net energy content value added and delivered to consumers is very small. And, the net benefit (all things considered) does little to reduce imported oil and fossil fuels required to produce the ethanol.

Although ethanol-from-corn and other food stocks has implications both in terms of world food prices and limited, yet positive energy yield (in terms of energy delivered to customer/fossil fuels used), the technology has led to the development of cellulosic ethanol. According to a joint research agenda conducted through the U.S. Department of Energy, the fossil energy ratios (FER) for cellulosic ethanol, corn ethanol, and gasoline are 10.3, 1.36, and 0.81, respectively.

Even dry ethanol has roughly one-third lower energy content per unit of volume compared to gasoline, so larger / heavier fuel tanks are required to travel the same distance, or more fuel stops are required. With large current unsustainable, non-scalable subsidies, ethanol fuel still costs much more per distance traveled than current high gasoline prices in the United States.

Methanol[22] is currently produced from natural gas, a non-renewable fossil fuel. It can also be produced from biomass as biomethanol[23]. The methanol economy is an interesting alternative to get to the hydrogen economy, compared to today's hydrogen production from natural gas. But this process is not the state-of-the-art clean solar thermal energy process where hydrogen production is directly produced from water.

<u>Butanol[24] is formed by ABE fermentation (acetone[25], butanol, ethanol) and experimental modifications of the process show potentially high net energy gains with butanol as the only liquid product. Butanol will produce more energy and allegedly can be burned "straight" in existing gasoline engines (without modification to the engine or car), and is less corrosive and less water soluble than ethanol, and could be distributed via existing infrastructures.</u> DuPont and BP are working together to help develop Butanol. *E. coli* have also been successfully engineered to produce Butanol by hijacking their amino acid metabolism.

Notes to the Difficult Sentences

Butanol is formed by ABE fermentation (acetone, butanol, ethanol) and experimental modifications of the process show potentially high net energy gains with butanol as the only liquid product. Butanol will produce more energy and allegedly can be burned "straight" in existing gasoline engines (without modification to the engine or car), and is less corrosive and less water soluble than ethanol, and could be distributed via existing infrastructures.

丁醇由 ABE 发酵（丙酮、丁醇和乙醇）产生，该过程在实验上经修改后可以获得高净能的丁醇作为唯一液体产物。丁醇将产生更高的能量并且据说能在现有的汽油发动机内直接燃烧（不需对发动机或车的结构进行修改），而且它比乙醇的腐蚀性更低且水溶性更小，能借助现有的基础设施进行推广。

Professional Words and Phrases

[1] **bioalcohol** [ˈbaiəuælkəhɔl] n. 生物醇（尤指乙醇）
[2] **ethanol** [ˈeθənəul] n. 乙醇
[3] **propanol** [ˈprəupənɔl] n. 丙醇
[4] **butanol** [ˈbju:tənəul] n. 丁醇
[5] **cellulose** [ˈseljuləus] n. 纤维素
[6] **biobutanol** [ˈbaibju:tənəul] n. 生物丁醇
[7] **biogasoline** [baiəuˈɡæsəli:n] n. 生化汽油
[8] **biodiesel** [ˈbaiəudi:zl] n. 生物柴油
[9] **bioethanol** [ˈbaiəueθənəul] n. 生物乙醇
[10] **ferment** [ˈfə:ment] v. 发酵
[11] **cellulosic** [seljuˈləusik] adj. 纤维素的
[12] **feedstock** [ˈfi:dstɔk] n. 原料
[13] **sugar beet** 糖用甜菜
[14] **sugar cane** 甘蔗
[15] **molasses** [məˈlæsiz] n. 糖浆，糖蜜
[16] **distillation** [distəlˈeiʃən] n. 蒸馏
[17] **bagasse** [bəˈɡæs] n. 甘蔗渣
[18] **octane** [ˈɔktein] n. 辛烷
[19] **thermal efficiency** 热效率
[20] **oxidizer** [ˈɔksədaizə] n. 氧化剂
[21] **fungicide** [ˈfʌndʒisaid] n. 杀真菌剂
[22] **methanol** [ˈmeθənəul] n. 甲醇
[23] **biomethanol** [ˈbaiəumeθənəul] n. 生物甲醇
[24] **butanol** [ˈbju:tənəul] n. 丁醇
[25] **acetone** [ˈæsətəun] n. 丙酮

Exercises

1. Matching

1) bioalcohol   a) a wide range of fuels which are in some way derived from biomass
2) biofuel   b) methanol produced from biomass
3) cellulose   c) an alcohol made by fermenting the sugar components of plant materials
4) bioethanol   d) an organic compound with the formula $(C_6H_{10}O_5)_n$, a polysaccharide consisting of a linear chain of several hundred to over ten thousand $\beta(1\rightarrow 4)$ linked D-glucose units
5) biomethanol   e) biologically produced alcohols

2. True or False

1) Ethanol can be used in petrol engines as a replacement for gasoline.
2) Ethanol has roughly the same energy content per unit of volume with gasoline.
3) Butanol will produce more energy and allegedly can be burned "straight" in existing gasoline engines.

3. Reading Comprehension

1) Which does bioalcohol refer to?
   A. bioethanol
   B. biobutanol
   C. biopropanol
   D. biomethanol

2) Which is not a product of ABE fermentation?
   A. acetone
   B. butanol
   C. methanol
   D. ethanol

3) Which is thought to be the best replacement of the gasoline?
   A. ethanol
   B. methanol
   C. butanol
   D. propanol

4. Translation from English to Chinese

Methanol is currently produced from natural gas, a non-renewable fossil fuel. It can also be produced from biomass as biomethanol. The methanol economy is an interesting alternative to get to the hydrogen economy, compared to today's hydrogen production from natural gas. But this process is not the state-of-the-art clean solar thermal energy process where hydrogen production is directly produced from water.

5. Translation from Chinese to English

生物产生的醇，多数情况下指的是乙醇，而很少指的是丙醇及丁醇，是由微生物和酶通过对糖、淀粉（最容易）或是纤维素（更难）的发酵产生。生物丁醇（也被称为生化汽油）经常被宣传可以直接替代汽油，因为它能直接用于汽油发动机（与生物柴油能用在柴油发动机类似）。

Chapter 10 Genomics

[本章中文导读]

基因组学是近些年来发展起来的研究生物基因组和如何利用基因的遗传学分支学科，涉及基因作图、测序和整个基因组功能分析。本章主要介绍最先测序的基因组（第10.1节）、功能基因组学（第10.2节）、蛋白质组学（第10.3节）以及生物信息学（第10.4节）等基因组学及其相关的基础专业知识。作者想通过利用英文讲解以上专业知识点，达到使学生熟悉并掌握一些基因组学的专业英语词汇的具体含义以及使用方法的目的。

10.1 The First Sequenced Genomes

The first genome to be sequenced, as you might expect, was a very simple one: The small DNA genome of an *E. coli* phage called φX174. Frederick Sanger, the inventor of the dideoxy chain termination method[1] of DNA sequencing[2], obtained the sequence of this 5375-nt genome in 1977.

What kind of information can we glean from this sequence? First, we can locate exactly the coding regions[3] for all genes and the distances between them to the exact nucleotide. How do we recognize a coding region? It contains an open reading frame[4] (ORF), a sequence of bases that, if translated in one frame, contains no stop codons for a relatively long distance-long enough to code for one of the phage proteins. Furthermore, the open reading frame must start with an ATG (or occasionally a GTG) triplet, corresponding to an AUG (or GUG) translation initiation codon[5], and end with a stop codon (UAG, UAA, or UGA). In other words, an open reading frame is the same as a gene's coding region.

The base sequence of the phage DNA also tells us the amino acid sequence of all the phage proteins. All we have to do is use the genetic code to translate the DNA base sequence of each open reading frame into the corresponding amino acid sequence. This may sound like a laborious process, but a personal computer can do it in a split second.

Sanger's analysis of the open reading frames of the φX174 DNA revealed something unexpected and fascinating: Some of the phage genes overlap. For example, the coding region for gene B lies within gene A and the coding region for gene E lies within gene D. Furthermore, genes D and J overlap by 1 bp. How can two genes occupy the same space and code for different proteins? The answer is that the two genes translated in different sets of codons will be encountered in these two frames, the two protein products will also be quite different.

This was certainly an interesting finding, and it raised the question of how common this phenomenon would be. So far, major overlaps seem to be confined almost exclusively to viruses, which is not surprising because these simple genetic systems[6] have small genomes in which the premium is on efficient use of the genetic material. Moreover, viruses have prodigious power to replicate, so enormous numbers of generations have passed during which evolution has honed the viral genomes.

With the advent of automated sequencing, geneticists have added much larger genomes to the list of total known sequences. In 1995, Craig Venter and Hamifton Smith and colleagues determined the entire base sequences of the genomes of two bacteria: *Haemophilus influenzae*[7] and *Mycoplasma genitalium*[8]. The *H. influenzae* (strain Rd) genome contains 1,830,137 bp and it was the first genome from a free-living organism to be completely sequenced. The *M. genitalrum* genome is the smallest of any known free-living organism and contains only about 470 genes. In April 1996, the leaders of an international consortium of laboratories announced another milestone: The 12-million-bp genome of baker's yeast (*Saccharomyces cerevrsiae*[9]) had been sequenced. This was the first eukaryotic genome to be entirely sequenced. Later in 1996, the first genome of an organism (*Methanococcus janaschii*[10]) from the third domain of life, the archaea, was sequenced. Then, in 1997, the long-awaited sequence of the 4.6 million-bp *E. coli* genome was reported. This is only about one-third the size of the yeast genome, but the importance of *E. coli* as a genetic tool made this a milestone as well. In 1998, the sequence of the first animal genome, from the roundworm *Caenorhabditis elegans*[11], was reported. The first plant genome (from the mustard family member *Arabidopsis thaliana*[12]) was completed in 2000. Also in 2000, the eagerly-awaited rough draft of the human genome sequence was announced. By 2001, this "working draft" of the human genome was published and dozens of bacterial genomes had been sequenced.

Notes to the Difficult Sentences

So far, major overlaps seem to be confined almost exclusively to viruses, which is not surprising because these simple genetic systems have small genomes in which the premium is on efficient use of the genetic material. Moreover, viruses have prodigious power to replicate, so enormous numbers of generations have passed during which evolution has honed the viral genomes.

到目前为止，主要的基因重叠看起来几乎毫无例外地存在于病毒内，这并不令人吃惊，因为这些最简单的遗传系统有非常小的基因组，而在基因组内最佳的方式就是有效地利用遗传物质。此外，病毒的复制能力是惊人的，所以在进化和优化病毒基因组的过程中病毒已经进行大量的传代。

Professional Words and Phrases

[1]　**dideoxy chain termination method**　双脱氧链终止法
[2]　**DNA sequencing**　DNA 测序
[3]　**coding region**　编码区
[4]　**open reading frame**　开放阅读框
[5]　**initiation codon**　起始密码子
[6]　**genetic system**　遗传系统
[7]　***Haemophilus influenzae***　流感嗜血杆菌

[8] *Mycoplasma genitalium*   生殖支原体
[9] *Saccharomyces cerevrsiae*   酿酒酵母
[10] *Methanococcus janaschii*   詹式甲烷球菌
[11] *Caenorhabditis elegans*   秀丽隐杆线虫
[12] *Arabidopsis thaliana*   拟南芥

Exercises

1. Matching

1) DNA sequencing
2) open reading frame
3) coding region
4) initiation codon
5) dideoxy chain termination method
6) genetic system

a) AUG (or GUG)
b) the methods and technologies that are used for determining the order of the nucleotide bases
c) the organization of genetic material and the ways in which the genetic material is transmitted
d) a DNA sequence that does not contain a stop codon in a given reading frame
e) portion of a gene's DNA or RNA, composed of exons, that codes for protein
f) DNA sequencing method based on the use of dideoxynucleotides (ddNTPs) in addition to the normal nucleotides (NTPs) found in DNA

2. True or False

1) The first genome to be completely sequenced is the DNA genome of an *E. coli* phage called φX174.
2) Major overlaps seem to be confined almost exclusively to viruses.
3) The *Saccharomyces cerevrsiae* genome was the first eukaryotic genome to be entirely sequenced.
4) The sequence of the first animal genome was reported later than that of the first plant genome.
5) The *M. genitalrum* genome is the smallest of any known free-living organism and contains only about 470 genes.

3. Reading Comprehension

1) Which is the first genome to be sequenced?
   A. *Haemophilus influenzae*
   B. φX174
   C. *Mycoplasma genitalium*
   D. *Methanococcus janaschii*

2) When was the E. coli genome sequenced?
   A. in 1996
   B. in 1997
   C. in 2000
   D. In 1998

3) Which is the first plant genome to be sequenced?

A. *Saccharomyces cerevrsiae*
B. *Methanococcus janaschii*
C. *Caenorhabditis elegans*
D. *Arabidopsis thaliana*

4. Translation from English to Chinese

Sanger's analysis of the open reading frames of the φX174 DNA revealed something unexpected and fascinating: Some of the phage genes overlap. For example, the coding region for gene B lies within gene A and the coding region for gene E lies within gene D. Furthermore, genes D and J overlap by 1 bp. How can two genes occupy the same space and code for different proteins? The answer is that the two genes translated in different sets of codons will be encountered in these two frames, the two protein products will also be quite different.

5. Translation from Chinese to English

噬菌体 DNA 的碱基序列也告诉了我们所有噬菌体蛋白质的氨基酸序列。我们所要做的是使用遗传密码将每个开放阅读框的 DNA 碱基序列翻译成相应的氨基酸序列。这可能听起来是个费劲的过程，但一台计算机在一秒内就能做到。

10.2　The Functional Genomics

Functional genomics[1] is a field of molecular biology that attempts to make use of the vast wealth of data produced by genomic projects (such as genome sequencing projects) to describe gene (and protein) functions and interactions. Unlike genomics and proteomics[2], functional genomics focuses on the dynamic aspects such as gene transcription, translation, and protein-protein interactions, as opposed to the static aspects of the genomic information such as DNA sequence or structures. Functional genomics attempts to answer questions about the function of DNA at the levels of genes, RNA transcripts, and protein products. A key characteristic of functional genomics studies is their genome-wide approach to these questions, generally involving high-throughput methods[3] rather than a more traditional "gene-by-gene" approach.

The goal of functional genomics is to understand the relationship between an organism's genome and its phenotype. The term functional genomics is often used broadly to refer to the many possible approaches to understanding the properties and function of the entirety of an organism's genes and gene products. This definition is somewhat variable; Gibson and Muse define it as "approaches under development to ascertain the biochemical, cellular, and/or physiological properties of each and every gene product", while Pevsner includes the study of nongenic elements[4] in his definition: "the genome-wide study of the function of DNA (including genes and nongenic elements), as well as the nucleic acid and protein products encoded by DNA". Because of its genome-wide approach, functional genomics requires the use of high-throughput technologies capable of assaying many functions or relationships simultaneously. Functional genomics involves studies of natural variation[5] in genes, RNA, and proteins over time (such as an organism's development) or space (such as its body regions), as well as studies of natural or experimental functional disruptions affecting genes, chromosomes, RNAs, or proteins.

The promise of functional genomics is to expand and synthesize genomic and proteomic knowledge into an understanding of the dynamic properties of an organism at cellular and/or organismal levels. This would provide a more complete picture of how biological function arises from the information encoded in an organism's genome. The possibility of understanding how a particular mutation leads to a given phenotype has important implications for human genetic diseases[6], as answering these questions could point scientists in the direction of a treatment or cure.

Functional genomics includes function-related aspects of the genome itself such as mutation and polymorphism (such as SNP) analysis, as well as measurement of molecular activities. The latter comprise a number of "-omics[7]" such as transcriptomics[8] (gene expression), proteomics (protein expression), phosphoproteomics[9] (a subset of proteomics) and metabolomics[10]. Functional genomics uses mostly multiplex techniques to measure the abundance of many or all gene products such as mRNAs or proteins within a biological sample. Together these measurement modalities endeavor to quantitate the various biological processes and improve our understanding of gene and protein functions and interactions.

Notes to the Difficult Sentences

The promise of functional genomics is to expand and synthesize genomic and proteomic knowledge into an understanding of the dynamic properties of an organism at cellular and/or organismal levels. This would provide a more complete picture of how biological function arises from the information encoded in an organism's genome. The possibility of understanding how a particular mutation leads to a given phenotype has important implications for human genetic diseases, as answering these questions could point scientists in the direction of a treatment or cure.

功能基因组学的目的是扩展并将基因组学和蛋白质组学知识组合起来用于理解生物在细胞和/或生物体水平上的动态特性。这将更全面地给人们提供一个对生物功能是怎样由生物基因组编码信息产生的概况。正如回答这些问题可以指引科学家治疗人的遗传疾病一样，理解一个特定突变怎样导致特定的表型可能会对人的遗传疾病具有重要的启示作用。

Professional Words and Phrases

[1]　**functional genomics**　功能基因组学
[2]　**proteomics**　['prəuti:əumiks]　n. 蛋白质组学
[3]　**high-throughput method**　高通量方法
[4]　**nongenic element**　非基因元件
[5]　**natural variation**　自然变异
[6]　**genetic disease**　遗传疾病
[7]　**omics**　['əumiks]　n. 生物组学
[8]　**transcriptomics**　['trænskriptəumiks]　n. 转录组学
[9]　**phosphoproteomics**　['fɔsfəuprəuti:əumiks]　n. 磷酸化转录组学
[10]　**metabolomics**　[metæ'bɔləmiks]　n. 代谢组学

Exercises

1. Matching

1) functional genomics
2) proteomics
3) transcriptomics
4) metabolomics
5) genomics

a) the scientific study of chemical processes involving metabolites
b) many possible approaches to understanding the properties and function of the entirety of an organism's genes and gene products
c) a discipline in genetics concerning the study of the genomes of organisms
d) the large-scale study of proteins, particularly their structures and functions
e) also referred to as expression profiling, the study of examination of the expression level of mRNAs in a given cell population

2. True or False

1) Functional genomics focuses on the static aspects such as gene transcription, translation, and protein-protein interactions.
2) The goal of functional genomics is to understand the relationship between an organism's genome and its phenotype.
3) Proteomics is used for the study of gene expression.

3. Reading Comprehension

1) Which is not included in the study of the functional genomics?
   A. gene transcription
   B. translation
   C. DNA sequence
   D. protein-protein interactions

2) Which is a subset of proteomics?
   A. genomics
   B. transcriptomics
   C. phosphoproteomics
   D. metabolomics

3) Which of the following statements is not true?
   A. Functional genomics is used for the study of natural variation in genes, RNA and proteins over time or space.
   B. Functional genomics includes function-related aspects of the genome itself and the measurement of molecular activities.
   C. Functional genomics uses mostly multiplex techniques to measure the abundance of many or all gene products within a biological sample.
   D. A traditional gene-by-gene approach is usually used for the study of functional genomics.

4. Translation from English to Chinese

Functional genomics includes function-related aspects of the genome itself such as mutation and polymorphism (such as SNP) analysis, as well as measurement of molecular activities. The latter comprise a number of "-omics" such as transcriptomics (gene expression), proteomics (protein expression), phosphoproteomics (a subset of proteomics) and metabolomics. Functional genomics uses mostly multiplex techniques to measure the abundance of many or all gene products such as mRNAs or proteins within a biological sample.

5. Translation from Chinese to English

与基因组学和蛋白质组学不同，功能基因组学关注诸如基因转录、翻译和蛋白质-蛋白质相互作用之类的动态方面的研究上，而不是 DNA 序列或结构等基因组信息的静态方面。功能基因组学试图回答 DNA 功能在基因、RNA 转录本和蛋白质产物水平上的有关问题。

10.3 Proteomics

Proteomics is the large-scale study of proteins, particularly their structures and functions. Proteins are vital parts of living organisms, as they are the main components of the physiological[1] metabolic pathways[2] of cells. The term "proteomics" was first coined in 1997 to make an analogy[3] with genomics, the study of the genes. The word "proteome" is a blend of "protein" and "genome", and was coined by Marc Wilkins in 1994 while working on the concept as a PhD student. The proteome[4] is the entire complement of proteins, including the modifications made to a particular set of proteins, produced by an organism or system. This will vary with time and distinct requirements, or stresses[5], that a cell or organism undergoes.

After genomics, proteomics is considered the next step in the study of biological systems. It is much more complicated than genomics mostly because while an organism's genome is more or less constant, the proteome differs from cell to cell and from time to time. This is because distinct genes are expressed in distinct cell types. This means that even the basic set of proteins which are produced in a cell needs to be determined.

In the past this was done by mRNA analysis[6], but this was found not to correlate with protein content[7]. It is now known that mRNA is not always translated into protein, and the amount of protein produced for a given amount of mRNA depends on the gene it is transcribed from and on the current physiological state of the cell. Proteomics confirms the presence of the protein and provides a direct measure of the quantity present.

Scientists are very interested in proteomics because it gives a much better understanding of an organism than genomics. First, the level of transcription of a gene gives only a rough estimate of its level of expression into a protein. An mRNA produced in abundance may be degraded rapidly or translated inefficiently, resulting in a small amount of protein. <u>Second, as mentioned above many proteins experience post-translational modifications that profoundly affect their activities; for example some proteins are not active until they become phosphorylated. Methods such as phosphoproteomics and glycoproteomics[8] are used to study post-translational modifications. Third, many transcripts give rise to more than one protein, through alternative splicing or alternative post-translational modifications. Fourth, many proteins form complexes with other proteins or RNA molecules, and only function in the presence of these other molecules.</u>

Finally, protein degradation rate[9] plays an important role in protein content.

One of the most promising developments to come from the study of human genes and proteins has been the identification of potential new drugs for the treatment of disease. This relies on genome and proteome information to identify proteins associated with a disease, which computer software can then use as targets for new drugs. For example, if a certain protein is implicated in a disease, its 3D structure provides the information to design drugs to interfere with the action of the protein. A molecule that fits the active site of an enzyme, but cannot be released by the enzyme, will inactivate the enzyme. This is the basis of new drug-discovery tools, which aim to find new drugs to inactivate proteins involved in disease. As genetic differences among individuals are found, researchers expect to use these techniques to develop personalized drugs that are more effective for the individual.

Notes to the Difficult Sentences

Second, as mentioned above many proteins experience post-translational modifications that profoundly affect their activities; for example some proteins are not active until they become phosphorylated. Methods such as phosphoproteomics and glycoproteomics are used to study post-translational modifications. Third, many transcripts give rise to more than one protein, through alternative splicing or alternative post-translational modifications. Fourth, many proteins form complexes with other proteins or RNA molecules, and only function in the presence of these other molecules. Finally, protein degradation rate plays an important role in protein content.

第二，正像上面提到的一样，许多蛋白质都将经历能深刻地影响其活性的翻译后修饰；例如一些蛋白质直到经过磷酸化才具有活性。磷酸化蛋白质组学和糖蛋白质组学等方法被用于翻译后修饰的研究。第三，许多转录本通过选择性剪切或选择性翻译后修饰产生不止一个蛋白。第四，许多蛋白质与其他蛋白质或 RNA 分子形成复合物，并且只有当这些其他分子存在时蛋白质才具有功能。最后，蛋白质降解率对蛋白质含量起到重要作用。

Professional Words and Phrases

[1] **physiological** [fiziəˈlɔdʒikəl] adj. 生理的，生理学的
[2] **metabolic pathway** 代谢途径
[3] **analogy** [əˈnælədʒi] n. 相似，类似
[4] **proteome** [ˈprəutiːəum] n. 蛋白质组
[5] **stress** [stres] n. 压力，胁迫
[6] **mRNA analysis** mRNA 分析
[7] **protein content** 蛋白质含量
[8] **glycoproteomics** [ˈglaikəuprəutiːəumiks] n. 糖蛋白质组学
[9] **protein degradation rate** 蛋白质降解率

Exercises

1. Matching

1) metabolic pathway   a) the entire set of proteins expressed by a genome, cell, tissue or

organism
- 2) glycoproteomics
- 3) phosphoproteomics
- 4) proteome
- 5) identification

b) an act of identifying

c) a branch of proteomics that identifies, catalogs, and characterizes proteins containing carbohydrates as a post-translational modification

d) a branch of proteomics that identifies, catalogs, and characterizes proteins containing a phosphate group as a post-translational modification

e) series of chemical reactions occurring within a cell

2. True or False

1) The word "proteome" is a blend of "protein" and "genome", and was coined by Marc Wilkins in 1994.

2) Proteomics gives a much better understanding of an organism than genomics.

3) The level of transcription of a gene is consistent with its level of expression into a protein.

3. Reading Comprehension

1) Which decides the amount of a protein in a cell?
   - A. the gene it is transcribed from
   - B. the current physiological state of the cell
   - C. both A and B
   - D. none of the above

2) Which of the following statements is not true?
   - A. Proteomics is much more complicated than genomics.
   - B. Many proteins experience post-translational modifications that profoundly affect their activities.
   - C. It is now known that mRNA is always translated into protein.
   - D. Many proteins form complexes with other proteins or RNA molecules, and only function in the presence of these other molecules.

3) Which is used to study the phosphorylated proteins?
   - A. phosphoproteomics
   - B. glycoproteomics
   - C. genomics
   - D. none of the above

4. Translation from English to Chinese

After genomics, proteomics is considered the next step in the study of biological systems. It is much more complicated than genomics mostly because while an organism's genome is more or less constant, the proteome differs from cell to cell and from time to time. This is because distinct genes are expressed in distinct cell types. This means that even the basic set of proteins which are produced in a cell needs to be determined.

5. Translation from Chinese to English

蛋白质组学进行的是对蛋白质尤其是其结构和功能的大规模的研究。因为蛋白质是细胞生理代谢途径的主要组分，所以它们是生物体的重要组成部分。专业术语"蛋白质组"是在1997年首先被创造出来的，用来与研究基因的基因组学进行类比研究。

10.4　Bioinformatics—Additional Reading

　　Bioinformatics is the application of statistics and computer science to the field of molecular biology. The term bioinformatics was coined by Paulien Hogeweg and Ben Hesper in 1978 for the study of informatic processes in biotic systems. Its primary use since at least the late 1980s has been in genomics and genetics, particularly in those areas of genomics involving large-scale DNA sequencing.

　　Bioinformatics now entails the creation and advancement of databases, algorithms, computational and statistical techniques and theory to solve formal and practical problems arising from the management and analysis of biological data.

　　Over the past few decades, rapid developments in genomic and other molecular research technologies and developments in information technologies have combined to produce a tremendous amount of information related to molecular biology. It is the name given to these mathematical and computing approaches used to glean understanding of biological processes.

　　Common activities in bioinformatics include mapping and analyzing DNA and protein sequences, aligning different DNA and protein sequences[1] to compare them and creating and viewing 3-D models of protein structures[2].

　　The primary goal of bioinformatics is to increase the understanding of biological processes. What sets it apart from other approaches, however, is its focus on developing and applying computationally intensive techniques (e.g., pattern recognition, data mining, machine learning algorithms, and visualization) to achieve this goal. Major research efforts in the field include sequence alignment[3], gene finding, genome assembly[4], drug design, drug discovery, protein structure alignment[5], protein structure prediction[6], prediction of gene expression and protein-protein interactions, genome-wide association studies and the modeling of evolution.

　　Bioinformatics was applied in the creation and maintenance of a database to store biological information at the beginning of the "genomic revolution", such as nucleotide and amino acid sequences. Development of this type of database involved not only design issues but the development of complex interfaces whereby researchers could both access existing data as well as submit new or revised data.

　　In order to study how normal cellular activities are altered in different disease states, the biological data must be combined to form a comprehensive picture of these activities. Therefore, the field of bioinformatics has evolved such that the most pressing task now involves the analysis and interpretation of various types of data, including nucleotide and amino acid sequences, protein domains[7], and protein structures. The actual process of analyzing and interpreting data is referred to as computational biology[8]. Important sub-disciplines within bioinformatics and computational biology include: ①the development and implementation of tools that enable efficient access to, and use and management of, various types of information. ②the development of new algorithms (mathematical formulas) and statistics with which to assess relationships among members of large data sets, such as methods to locate a gene within a sequence, predict protein structure and/or function, and cluster protein sequences into families of related sequences.

Notes to the Difficult Sentences

　　Important sub-disciplines within bioinformatics and computational biology include: ①the development and implementation of tools that enable efficient access to, and use and management of, various types of information. ②the development of new algorithms (mathematical formulas) and statistics with which to assess relationships among members of large data sets, such as methods to locate a gene within a sequence, predict protein structure and/or function, and cluster protein sequences into families of related sequences.

　　重要的生物信息学和计算生物学子学科包括：①运用和发展能有效获得、使用和管理各类信息的工具。②发展能用以评价大型数据集成员之间相互关系的新算法（数学公式）和统计学方法，例如，对基因在序列中的位置进行定位、对蛋白质结构和/或功能进行预测、将蛋白质序列归入与其相关的序列家族等方法。

Professional Words and Phrases

[1] **protein sequence** 蛋白质序列
[2] **protein structure** 蛋白质结构
[3] **sequence alignment** 序列比对
[4] **genome assembly** 基因组拼接
[5] **protein structure alignment** 蛋白质结构比对
[6] **protein structure prediction** 蛋白质结构预测
[7] **protein domain** 蛋白质结构域
[8] **computational biology** 计算生物学

Exercises

1. Matching

1) bioinformatics　　a) the process of taking a large number of short DNA sequences, all of which were generated by a shotgun sequencing project, and putting them back together to create a representation of the original chromosomes from which the DNA originated

2) sequence alignment　　b) The actual process of analyzing and interpreting biological data

3) protein domain　　c) a way of arranging the sequences of DNA, RNA, or protein to identify regions of similarity

4) computational biology　　d) a part of protein sequence and structure that can evolve, function, and exist independently of the rest of the protein chain

5) genome assembly　　e) the application of statistics and computer science to the field of molecular biology

2. True or False

1) Bioinformatics is the application of statistics and computer science to the field of molecular biology.

2) The term bioinformatics was coined by Paulien Hogeweg and Ben Hesper in the late 1980s for the study of informatic processes in biotic systems.

3) Bioinformatics now entails the creation and advancement of databases, algorithms, computational and statistical techniques and theory to solve formal and practical problems arising from the management and analysis of biological data.

3. Reading Comprehension

1) Which is included in the common activities of bioinformatics?
   A. mapping and analyzing DNA and protein sequences
   B. aligning different DNA and protein sequences to compare them
   C. creating and viewing 3-D models of protein structures
   D. all

2) Which of the following statements is not true?
   A. The primary goal of bioinformatics is to increase the understanding of biological processes.
   B. Bioinformatics is mainly used in the genomics and genetics now.
   C. Bioinformatics is now used in the creation and maintenance of a database to store biological information.
   D. The main difference between bioinformatics and other approaches is developing and applying computationally intensive techniques to increase the understanding of biological processes.

3) Which data are studied in bioinformatics?
   A. nucleotide and amino acid sequence
   B. protein domains
   C. protein structures
   D. all

4. Translation from English to Chinese

The primary goal of bioinformatics is to increase the understanding of biological processes. What sets it apart from other approaches, however, is its focus on developing and applying computationally intensive techniques (e.g., pattern recognition, data mining, machine learning algorithms, and visualization) to achieve this goal. Major research efforts in the field include sequence alignment, gene finding, genome assembly, drug design, drug discovery, protein structure alignment, protein structure prediction, prediction of gene expression and protein-protein interactions, genome-wide association studies and the modeling of evolution.

5. Translation from Chinese to English

生物信息学在"基因组革命"之初主要是被应用于创建和维护存储诸如核苷酸序列和氨基酸序列等生物信息的数据库。这类数据库的开发不仅涉及设计问题，还涉及复杂接口的开发，使研究人员既可以访问现有数据，也可以提交新的或修订的数据。

Appendix I Answers to Exercises

Chapter 1

1.1 What is Biology?

Exercises

1. Matching
   d a g h b e i c f
2. True or False
   F T F F T
3. Reading Comprehension
   D A A
4. Translation from English to Chinese
   　　近些年来，生物学的成就集中在致力于基因组测序和序列比对的基因组学上，以及旨在利用事先编写好的 DNA 程序创造生命的合成生物学上。
5. Translation from Chinese to English
   　　Life, while complex, consists of a finite amount of complexity that only appreciably increases on relatively long timescales of hundreds of thousands or millions of years. Evolution, while creative, operates slowly.

1.2 The Origin of Life

Exercises

1. Matching
   d c a e b
2. True or False
   F F F F T
3. Reading Comprehension
   C D D
4. Translation from English to Chinese
   　　UCLA（美国加州大学洛杉矶分校）的古生物学家 William Schopf 在 2002 年发表的一篇有争议的论文争论说叠层石的波浪形地质构造实际上含有 35 亿年前的藻类（一种微生物）化石。一些古生物学家则不赞同 Schopf 的结论，他们预计最早的生命形式开始于 30 亿年前左右而不是 35 亿年前。
5. Translation from Chinese to English
   　　If certain microspheres formed which were capable of encouraging the growth of additional microspheres around them, it would amount to a primitive form of self-replication, and eventually Darwinian evolution would take over, creating effective self-replicators like today's cyanobacteria.

1.3 The Significance of Biology in Your Life

Exercises

1. Matching
   c e a b f d
2. True or False
   T F F F F
3. Reading Comprehension
   D A D

4. Translation from English to Chinese

　　生物学在日常生活中的重要性体现在它努力寻找存在于具有各种形态和功能的生物体中的统一规律。生物学在日常生活中的重要性可以从科学本身的两个自然门类——植物学和动物学来进行理解。

5. Translation from Chinese to English

　　Agriculture is largely the result of man's taking the advantage of the interrelations of soil, climate and natural habitat to select those particular combinations that meet his basic requirements. Thus to provide necessary food, man depends entirely on green plants that can alone capture the solar energy. High yielding varieties of crop plants like rice, wheat, jute, sugar cane, pulses etc are now bred experimentally. Disease-resistant grains and vernalized seeds are made.

1.4　The History of Biology—Additional Reading

Exercises

1. Matching

　　g e d c f a b

2. True or False

　　F F T F F

3. Reading Comprehension

　　D C D

4. Translation from English to Chinese

　　新学科得以快速发展，尤其是在 Waston（沃森）和 Crick（克里克）提出 DNA 结构之后。随着"中心法则"的建立和遗传密码的破解，生物学主要分成了处理整个生物体和生物群落的领域即有机体生物学和与细胞和分子生物学相关的领域两部分。二十世纪后期，像基因组学和蛋白质组学等新领域正在颠覆这一趋势，有机体生物学家开始使用分子技术，而分子和细胞生物学家开始研究基因与环境的相互影响以及自然生物群落的遗传学。

5. Translation from Chinese to English

　　Advances were made in analytical chemistry and physics instrumentation including improved sensors, optics, tracers, instrumentation, signal processing, networks, robots, satellites, and compute power for data collection, storage, analysis, modeling, visualization, and simulations. These technology advances allowed theoretical and experimental research including internet publication of molecular biochemistry, biological systems, and ecosystems science.

Chapter 2　Microbiology

2.1　The Scope and relevance of Microbiology

Exercises

1. Matching

　　c d e a h j i b f g

2. True or False

　　T F T T F T F

3. Reading Comprehension

　　D B D D B

4. Translation from English to Chinese

　　正如科学家、作家 Steven Jay Goul 强调的那样，我们处在细菌的年代。它们是我们这个星球最早的生命体，生活在可能存在生命的任何地方，在数量上比其他任何生命都多，并且可能是地球生物量最大的组成部分。整个生态系统依赖于它们的活动，并且以无数的方式影响着人类社会。因此，现代微生物学是一门包含许多不同专业的大学科；它对诸如医学、农业和食品科学、生态学、遗传学、生物化学和分子生物学等都具有很大的影响力。

5. Translation from Chinese to English

   Research on the biology of microorganisms occupies the time of many microbiologists and also has practical applications. Those working in microbial physiology and biochemistry study the synthesis of antibiotics and toxins, microbial energy production, the ways in which microorganisms survive harsh environmental conditions, microbial nitrogen fixation, the effects of chemical and physical agents on microbial growth and survival, and many other topics.

2.2 The Future of Microbiology

Exercises

1. Matching

   g c e f b d a

2. True or False

   T F F T T T F

3. Reading Comprehension

   D A D

4. Translation from English to Chinese

   微生物对工业和环境控制越来越重要，并且我们不得不了解各种利用它们的新方法。例如，微生物能（a）用于生产高质量食品和其他能应用在工业上的酶等应用产物，（b）用于降解污染物和有毒物质，（c）作为治疗疾病和提高农业生产力的载体。此外，食品和农作物需要得到持续保护以免受微生物的损害。

5. Translation from Chinese to English

   Further research on unusual microorganisms and microbial ecology will lead to a better understanding of the interactions between microorganisms and the inanimate world. Among other things, this understanding should enable us to more effectively control pollution. Similarly, it has become clear that microorganisms are essential partners with higher organisms in symbiotic relationships. Greater knowledge of symbiotic relationships can help improve our appreciation of the living world. It also will lead to improvements in the health of plants, livestock, and humans.

2.3 Prokaryotes, Eukaryotic Microbes and Viruses

Exercises

1. Matching

   e d b f g i h c a

2. True or False

   F F T F T

3. Reading Comprehension

   D B C

4. Translation from English to Chinese

   显然病毒与原核微生物和真核微生物是截然不同的，它们是由病毒学家负责研究的。尽管病毒比由细胞组成的生命体更简单，但是它们是极为重要的，应予以高度重视。病毒学曾对分子生物学做出了突出的贡献。许多人类病毒类疾病已经是众所周知的，而且每年更多的此类疾病还在不断地被发现。

5. Translation from Chinese to English

   The classification of eukaryotic microbes is problematic and has changed frequently. Historical schemes based on similarity in morphology and chemistry have been replaced with schemes based on nucleotide sequences and ultrastructural features. There are three major eukaryotic microorganisms: fungi, algae and protozoa.

2.4 Extreme Microbes—Additional Reading

Exercises

1. Matching

   d e g f c b a

2. True or False

 F F T T F

3. Reading Comprehension

 D B C

4. Translation from English to Chinese

最先发现的极端微生物之一是在科学领域非常著名的水生栖热菌，它是最先在黄石国家公园的天然热喷泉中发现的。这一细菌能在 66℃时生存并进行繁殖。该温度是实验室中复制 DNA 的理想温度，因为热能加速反应进行使得反应更快地完成。结果，聪明的科学家就将该菌的 *Taq* 聚合酶这一关键的复制 DNA 的蛋白质应用于实验室。今天 *Taq* 已经在世界范围内广泛被应用于生物实验室，并且它是使得法医的 DNA 分析鉴定成为可能的原因之一。

5. Translation from Chinese to English

Other microbes were recently discovered that actually eat the byproducts of radiation. Scientists found a species of bacteria in South Africa living two miles (1mile=1609 meters) below Earth's surface in high-pressure environments. These bacteria, existing without sun and all other forms of life, and which may have been lurking in those depths for up to 25 million years, get their nourishment from radiation byproducts produced by the reaction of uranium, thorium and potassium with water over time. They might be the oldest ancestors of current bacteria, raising the question of whether life actually began underground.

Chapter 3　Cellular Biology

3.1　The Discovery of Cells

Exercises

1. Matching

 e d f b a g c

2. True or False

 T F F

3. Reading Comprehension

 D A C

4. Translation from English to Chinese

直到 19 世纪 30 年代，细胞的重要性才广泛达成共识。在 1838 年，一个以前是德国律师的植物学家 Matttuas Schleiden 断定植物的组织尽管各不相同，但都是由细胞组成的，而且植物胚胎来自单细胞。在 1839 年，一个德国的动物学家（Schleiden 实验室的同事）Theodor Schwann 发表了一份有关于动物生命细胞基础的综合报告。Schwann 断定植物和动物细胞结构是相似的，并提出了细胞理论的两个重要原则：①所有的生命体都是由一个或多个细胞组成的；②细胞是生命结构单位。

5. Translation from Chinese to English

Schleiden and Schwann's ideas on the *origin of cells* proved to be less insightful; both agreed that cells could arise from noncellular materials. Given the prominence that these two scientists held in the scientific world, it took a number of years before observations by other biologists were accepted as demonstrating that cells did not arise in this manner any more than organisms arose by spontaneous generation. By 1855, Rudolf Virchow, a German pathologist, had made a convincing case for the third tenet of the cell theory: Cells can arise only by division from a preexisting cell.

3.2　Basic Properties of Cells

Exercises

1. Matching

 g h d i a b c k m l e j f

2. True or False

 T T F T F

3. Reading Comprehension

C D C

4. Translation from English to Chinese

我们正逐步地增进对细胞控制其活动方式的了解，但是还有更多的调控方式有待发现。在细胞中，控制产物的信息存在于核酸，而构建产物的"工人"则主要为蛋白质。这两类大分子（而不是其他任何因子）的存在使得细胞的化学性质有别于非生命界的化学性质。在细胞中，"工人"必须按照自觉有序的方向进行工作。每一步过程必须以自动开启下一步的方式自然发生。所有需要指引的特定活动——无论是蛋白质合成、激素分泌，还是肌肉纤维收缩的信息都必须存在于系统本身之内。

5. Translation from Chinese to English

Some cells respond to stimuli in obvious ways; a single-celled protist, for example, moves away from an object in its path or moves toward a source of nutrients. Cells within a multicellular plant or animal respond to stimuli less obviously, but they respond, nonetheless. Most cells are covered with receptors that interact with substances in the environment in highly specific ways. Cells possess receptors to hormones, growth factors, extracellular materials, as well as to substances on the surfaces of other cells. A cell's receptors provide doorways through which external agents can evoke specific responses in target cells. Cells may respond to specific stimuli by altering their metabolic activities, preparing for cell division, moving from one place to another, or even committing suicide.

3.3 The Concepts in Mammalian Cell Culture—Additional Reading

Exercises

1. Matching

d g h a f e i j b c

2. True or False

T T T F F

3. Reading Comprehension

A C C

4. Translation from English to Chinese

细胞能以悬浮或贴壁的方式生长在培养物中。有些细胞在没有贴壁的表面存在时可以自然地以悬浮方式生活，例如在血液中的细胞。也有细胞系经修饰能以悬浮的方式生存于培养物中，以至于它们可以生长到比贴壁条件培养所允许的密度更高的密度。贴壁细胞需要用于组织培养的塑料制品或微载体等的一个表面，这一表面可以用细胞外基质组分进行包裹，这样可以增加其黏附能力并可以为细胞提供其生长和分化所需要的其他信号。多数来源于固体组织的细胞是具有黏附能力的。还有一种贴壁培养类型是组织型培养，它是在三维环境中培养细胞，与二维的培养皿不同。

5. Translation from Chinese to English

Another common method for manipulating cells involves the introduction of foreign DNA by transfection. This is often performed to cause cells to express a protein of interest. More recently, the transfection of RNAi constructs have been realized as a convenient mechanism for suppressing the expression of a particular gene/protein. DNA can also be inserted into cells using viruses, in methods referred to as transduction, infection or transformation. Viruses, as parasitic agents, are well suited to introducing DNA into cells, as this is a part of their normal course of reproduction.

Chapter 4   Botany—Plant Biology

4.1   The Scope and Importance of Botany

Exercises

1. Matching

c d a f g e b

2. True or False

　T T F

3. Reading Comprehension

　D A C

4. Translation from English to Chinese

　　　植物学也称植物科学或植物生物学，是生物学的一个分支，涉及植物生命相关的科学研究。植物学涵盖了涉及植物、藻类和真菌研究的广泛的学科，所研究的是各个分类群体的结构、生长、生殖、代谢、发育、疾病、化学特性以及进化关系。植物学早期是从人类致力于鉴别什么是可食的、药用的以及有毒的植物开始的，这使得它成了一门最古老的科学。

5. Translation from Chinese to English

　　　The study of plants is vital because they are a fundamental part of life on Earth, which generates the oxygen, food, fibers, fuel and medicine that allow humans and other life forms to exist. Through photosynthesis, plants absorb carbon dioxide, a greenhouse gas that in large amounts can affect global climate. Additionally, they prevent soil erosion and are influential in the water cycle. A good understanding of plants is crucial to the future of human societies as it allows us to: ①Produce food to feed an expanding population. ②Understand fundamental life processes. ③Produce medicine and materials to treat diseases and other ailments. ④Understand environmental changes more clearly.

4.2　Flowers, Fruits and Seeds of Plants

Exercises

1. Matching

　c d a f e g b

2. True or False

　T F F F F

3. Reading Comprehension

　A C D

4. Translation from English to Chinese

　　　果皮是果实的一部分，是子房壁成熟后形成的。它由三层组成：表皮（外果皮）、通常是肉质状的中部（中果皮）以及薄膜状的或像石头一样坚硬的内层（内果皮）。这些分层可能在果实形成期间变成像皮肤似的、皮质化的、肉质化的或是纤维（硬化）的。果实产生保护和传播种子的装置。此外，花其他部分也可以对果实的形成起作用，比如花轴和花萼。

5. Translation from Chinese to English

　　　Numerous devices support the dispersal of the descendants since only a suitable habitat enables the seed to develop into a viable plant. If the dispersal is achieved only through devices that the plant itself produces, it is spoken of self-dispersal or autochory. If extern factors such as wind, water, animals, etc. are involved, then the mode of dispersal is called allochor. Seeds are dispersed either by self-dispersal, also called autochory, or by allochory which means that extern factors are involved. Allochor modes of dispersal are dispersal by wind (anemochory), water (hydrochory) or by animals (zoochory).

4.3　Photosynthesis

Exercises

1. Matching

　f g a e b h d c

2. True or False

　F F T T T

3. Reading Comprehension

C A C

4. Translation from English to Chinese

光反应发生在类囊体膜上，并将光能转化成化学能。因此，化学反应必须发生在光存在时。叶绿素和β-胡萝卜素等其他几种色素成簇地结合在类囊体膜内，并参与光反应。这些不同颜色的色素能吸收颜色略微不同的光，并将其能量传送到处于反应中心的叶绿素分子进行光合作用。

5. Translation from Chinese to English

The dark reaction takes place in the stroma within the chloroplast, and converts CO_2 to sugar. This reaction doesn't directly need light in order to occur, but it does need the products of the light reaction (ATP and another chemical called NADPH). The dark reaction involves a cycle called the Calvin cycle in which CO_2 and energy from ATP are used to form sugar.

4.4 Some Achievements of Transgenic Plants—Additional Reading

Exercises

1. Matching

d a g f c e b

2. True or False

T F T T F

3. Reading Comprehension

A B A

4. Translation from English to Chinese

苏云金芽孢杆菌是对许多昆虫类害虫具有致病力的一种细菌。其致命作用是由其自身产生的一种蛋白质介导的。通过重组 DNA 技术，毒素基因能被直接引入植物基因组内，并在那得以表达并保护植物免受昆虫类害虫的侵害。

5. Translation from Chinese to English

A large fraction of the world's irrigated crop land is so laden with salt that it cannot be used to grow most important crops. However, researchers at the University of California Davis campus have created transgenic tomatoes that grew well in saline soils. The transgene was a highly-expressed sodium/proton antiport pump that sequestered excess sodium in the vacuole of leaf cells. There was no sodium buildup in the fruit.

Chapter 5   Zoology – Animal Biology

5.1   What is Zoology?

Exercises

1. Matching

c a b

2. True or False

T F T T T

3. Reading Comprehension

C B B

4. Translation from English to Chinese

现代动物学是随着显微镜的发明到来的，但忽略 Charles Darwin（达尔文）极为重要的贡献是不合适的。正是由于达尔文的进化论，人类才得以不断增加对动物和人类之间的关系的理解，并进一步激发了人类对动物的兴趣，这是因为人类曾被一些人认为是动物的后代。当达尔文理论得到接受时，人类才被认为是作为动物界的一部分被包含在其内，尽管这一共识本身没有歧视性但仍让人类感到羞辱。

5. Translation from Chinese to English

Zoology is a branch of biology that focuses on the study of animals. Within this branch, people may also specialize and study certain forms of animals. For example a person working in zoology might study fish biology and work as an ichthyologist. Alternately, a zoologist could specialize in the study of mammals and be called a mammalogist.

5.2 Tissues, Organs and Organ Systems of Animals

Exercises

1. Matching

   f h e g b d a c

2. True or False

   T F T F T

3. Reading Comprehension

   A D C

4. Translation from English to Chinese

在哺乳动物中，组织来源于胚胎的三个分层。外胚层（外层）产生皮肤和神经组织；中胚层（中层）产生肌肉、骨骼，以及许多生殖、泌尿和循环器官；内胚层（内层）产生消化道的内侧和肺等器官。组织由细胞和细胞外产物组成。器官由数种组织组成，而器官系统由数个器官组成。器官系统执行的是消化、交流、循环、呼吸、分泌和运动等机体功能。

5. Translation from Chinese to English

Tissues may be categorized into four major types: ①epithelial, ②connective tissue, ③muscle, and ④nervous tissue. The cells of tissues are held together by one or more of a variety of cell junctions. Some are tight junctions which do not let fluids pass, some are composed of supporting filaments to give the cell shape and to attach the cell to its neighboring cells (adhering junctions), and some, such as the gap junctions, are for intercellular communication.

5.3 Transgenic Animals—Additional Reading

Exercises

1. Matching

   d f e a b g c

2. True or False

   T T T

3. Reading Comprehension

   A A C

4. Translation from English to Chinese

生产转基因动物的基本原理是将一个或数个外源基因引入动物（插入的基因称输入基因）。外源基因"必须通过生殖系被转移，这样包括生殖细胞在内的每个动物细胞含有一样的经修饰的遗传物质"。（生殖细胞是能将基因转移到生命体后代的细胞。）目前为止，产生转基因动物的基本方法有三类：①DNA显微注射，②逆转录病毒介导的基因转移，③胚胎干细胞介导的基因转移。

5. Translation from Chinese to English

Gene transfer by microinjection is the predominant method used to produce transgenic farm animals. Since the insertion of DNA results in a random process, transgenic animals are mated to ensure that their offspring acquire the desired transgene. However, the success rate of producing transgenic animals individually by these methods is very low and it may be more efficient to use cloning techniques to increase their numbers.

Chapter 6 Molecular Genetics

6.1 Gene and Chromosomes

Exercises

1. Matching

   f d e c b a

2. True or False

   T F F F F T T

3. Reading Comprehension

   C B B

4. Translation from English to Chinese

   原核生物——细菌和古菌以一种典型的方式将基因组储存在一个单一的、大的、环状的染色体上，额外的称为质粒的小环 DNA 有时是基因组在染色体外的补充部分，它通常编码一小部分基因并且容易在个体之间进行转移。例如，抗生素抗性基因通常在质粒上，并且可以在个体细胞之间，甚至是不同物种之间，借助水平基因转移进行传递。

5. Translation from Chinese to English

   Whereas the chromosomes of prokaryotes are relatively gene-dense, those of eukaryotes often contain so-called "junk DNA", or regions of DNA that serve no obvious function. Simple single-celled eukaryotes have relatively small amounts of such DNA, whereas the genomes of complex multicellular organisms, including humans, contain an absolute majority of DNA without an identified function. However it now appears that, although protein-coding DNA makes up barely 2% of the human genome, about 80% of the bases in the genome may be expressed, so the term "junk DNA" may be a misnomer.

6.2 Functional Structure of a Gene

Exercises

1. Matching

   f d e b g c a

2. True or False

   F F T T F

3. Reading Comprehension

   D C A

4. Translation from English to Chinese

   在这个结构中，碱基配对原则指定鸟嘌呤与胞嘧啶配对，而腺嘌呤与胸腺嘧啶配对。鸟嘌呤与胞嘧啶之间的碱基配对形成三条氢键，而腺嘌呤与胸腺嘧啶之间的碱基配对则形成两条氢键。因此，双螺旋的两条链必须是互补的，换句话说，碱基必须以此种方式排列才能使得一条链的腺嘌呤与另一条链的胸腺嘧啶配对，另外两种碱基也是如此。

5. Translation from Chinese to English

   A gene can have more than one promoter, resulting in RNAs that differ in how far they extend in the 5' end. Although promoter regions have a consensus sequence that is the most common sequence at this position, some genes have "strong" promoters that bind the transcription machinery well, and others have "weak" promoters that bind poorly. These weak promoters usually permit a lower rate of transcription than the strong promoters, because the transcription machinery binds to them and initiates transcription less frequently. Other possible regulatory regions include enhancers, which can compensate for a weak promoter.

6.3 Gene Expression

Exercises

1. Matching

   e c d b a

2. True or False

   T F F

3. Reading Comprehension

   A C C

4. Translation from English to Chinese

   转录是通过一种称为 RNA 聚合酶的酶进行的,该酶从 3'端向 5'端读取模板链,而从 5'端向 3'端合成 RNA。为了使转录起始,该聚合酶先识别并结合基因的一个启动子区域。这样,基因调控的主要机制就是阻遏物分子通过与启动子区紧密结合才能阻止聚合酶的结合,或是通过改变 DNA 的结构使得聚合酶无法接近启动子区。

5. Translation from Chinese to English

   In all organisms, there are two major steps separating a protein-coding gene from its protein: First, the DNA on which the gene resides must be transcribed from DNA to messenger RNA (mRNA); and, second, it must be translated from mRNA to protein. RNA-coding genes must still go through the first step, but are not translated into protein. The process of producing a biologically functional molecule of either RNA or protein is called gene expression, and the resulting molecule itself is called a gene product.

6.4 Essentials for Genetic Engineering—Additional Reading

Exercises

1. Matching

   e f g a c d b

2. True or False

   T F T F F

3. Reading Comprehension

   D B D

4. Translation from English to Chinese

   电转化就是这样一种技术。质粒越大,被细胞吸收的效率就越低。越大的 DNA 片段越容易使用噬菌体、逆转录病毒、其他的病毒载体或黏粒进行克隆,所使用的方法称为转导。噬菌体或病毒载体经常应用于再生医学领域,但是这可能会导致 DNA 插入我们并不想要其插入的部分染色体区域,进而导致并发症甚至癌症。

5. Translation from Chinese to English

   Gene cloning is the act of making copies of a single gene. Once a gene is identified, clones can be used in many areas of biomedical and industrial research. Genetic engineering is the process of cloning genes into new organisms, or altering the DNA sequence to change the protein product. Genetic engineering depends on our ability to perform the following essential procedures.

Chapter 7 Biochemistry

7.1 Enzymes

Exercises

1. Matching

   e c d a b

2. True or False

   T F T

3. Reading Comprehension

   C C C

4. Translation from English to Chinese

   诱导契合假说暗示底物与酶结合会改变酶的结构，进而使得底物与底物之间结合更紧密而进一步使反应更容易发生。酶的辅因子是酶反应必需的非蛋白质成分。像 K^+ 和 Ca^{2+} 等离子就是辅因子。与之不同的是，辅酶是结合在活性部位附近的非蛋白质有机分子。

5. Translation from Chinese to English

   Enzymes are proteins. The functioning of the enzyme is determined by the shape of the protein. The arrangement of molecules on the enzyme produces an area known as the active site within which the specific substrate(s) will "fit". It recognizes, confines and orients the substrate in a particular direction.

7.2 Metabolism

Exercises

1. Matching

   e c a b d

2. True or False

   F T T

3. Reading Comprehension

   D C D

4. Translation from English to Chinese

   一个分解代谢的例子是细胞从环境中吸收葡萄糖分子，将其分解并释放能量（即糖酵解）。能量释放后立即被一个特殊分子（腺苷三磷酸或称 ATP）捕捉并储存。当储存在 ATP 中的能量用于合成更复杂的化合物（例如，数个单糖结合形成二糖或多糖）时，该过程被称为合成代谢。

5. Translation from Chinese to English

   Catabolism means disintegration, whereas anabolism means reorganization. The catabolism implies release of energy, whereas anabolism implies capture of energy. Catabolism implies disorganization of matter, whereas anabolism implies a more complex reorganization of matter.

7.3 Energy Transformation

Exercises

1. Matching

   d e a c b

2. True or False

   T T F

3. Reading Comprehension

   C D C

4. Translation from English to Chinese

   氨基酸可以被用于合成蛋白质和其他生物分子，也可被用作能量来源被氧化成尿素和二氧化碳。氧化途径开始于转氨酶移走氨基酸的氨基。氨基进入尿素循环，而脱氨基剩下的碳骨架形成酮酸。这些酮酸中的几个是三羧酸循环中的中间产物，例如，谷氨酸脱氨形成 α-酮戊二酸。生糖氨基酸也可以通过糖异生作用转换成葡萄糖。

5. Translation from Chinese to English

   Chemolithotrophy is a type of metabolism found in prokaryotes where energy is obtained from the oxidation of inorganic compounds. These organisms can use hydrogen, reduced sulfur compounds (such as sulfide, hydrogen sulfide and thiosulfate), ferrous iron (FeⅡ) or ammonia as sources of reducing power and they gain energy from the oxidation of these compounds with electron acceptors such as oxygen or nitrite. These microbial processes are important in global biogeochemical cycles such as acetogenesis, nitrification and denitrification and are critical for soil fertility.

7.4 Protein crystallization—Additional Reading
Exercises
1. Matching

　　d e a c b

2. True or False

　　T F F T F

3. Reading Comprehension

　　B C D

4. Translation from English to Chinese

　　结晶的目的是产生一个排列非常好并且含污染物少的晶体，当其足够大就可以用作暴露在 X 射线下的衍射模型。然后该衍射模型可以被分析，用以识别蛋白质的三级结构。蛋白质结晶本来就是很困难的，这是由于蛋白质晶体本身的脆弱性。蛋白质有不规则的表面，这导致在任何蛋白质晶体中可以形成大的孔道。因此，将晶格组合在一起的非共价键通常是必须通过几层溶剂分子才能形成的。

5. Translation from Chinese to English

　　In order to crystallize a protein, the purified protein undergoes slow precipitation from an aqueous solution. As a result, individual protein molecules align themselves in a repeating series of unit cells by adopting a consistent orientation. The crystalline lattice that forms is held together by noncovalent interactions. The importance of protein crystallization is that it serves as the basis for X-ray crystallography, wherein a crystallized protein is used to determine the protein's three-dimensional structure via X-ray diffraction.

Chapter 8　Ecology
8.1　What is Ecology?
Exercises
1. Matching

　　c e a b d

2. True or False

　　F T T

3. Reading Comprehension

　　B D A

4. Translation from English to Chinese

　　生态学与环境、环保论、自然历史学或环境科学并不是一个概念。它与生理学、进化生物学、遗传学和动物行为学等则是密切相关的。理解生物的多样性如何影响生态功能是生态学研究中的一个重点领域。生态系统维系着这个星球上支撑每个生命的功能，这包括气候调节、水体过滤、土壤形成（土壤发生）、食物、纤维、医药、侵蚀控制以及许多具有科学、历史或精神价值的其他自然特征。

5. Translation from Chinese to English

　　Ecologists seek to explain: ①life processes and adaptations, ②distribution and abundance of organisms, ③the movement of materials and energy through living communities, ④the successional development of ecosystems, ⑤the abundance and distribution of biodiversity in context of the environment.

8.2 Ecosystems
Exercises

1. Matching

    e a d b c

2. True or False

    T T F

3. Reading Comprehension

    D B C

4. Translation from English to Chinese

    生态动力学的范畴可以视作像蚜虫在树上迁移这样的局部位置这一变量的封闭体系，而同时对像大气或气候这样的更大范围的影响则仍是开放的。因此，生态学家可以通过分析对从像植物群丛、气候和土壤类型这样的较细微尺度单元收集来的数据设计生态系统的等级结构分类方法，并整合这些信息以鉴定在地域、局部和时间尺度上运行且组成和过程均一的更大突发模式。

5. Translation from Chinese to English

    Ecosystems are confronted with natural environmental variations both over time and by differences between locations. The variations occur over vastly different ranges in terms of magnitudes as well as distances and time periods. It can take thousands of years for ecological processes to mature, for example, for all the successional stages of a forest. The area of an ecosystem can vary greatly from tiny to vast.

8.3 Biodiveristy—Additional Reading
Exercises

1. Matching

    b a e f c d

2. True or False

    T F T F F

3. Reading Comprehension

    D C B

4. Translation from English to Chinese

    单词"生物多样性"是"生物学的多样性"的缩写。《生物多样性公约》则将生物多样性定义为："其中包括陆地、海洋及其他水生系统在内的所有来源的生命体的变异性，以及生态系统作为其一部分的生态学复合体；这包括物种内、物种间以及生态系统的多样性。"

5. Translation from Chinese to English

    Biological diversity is of fundamental importance to the functioning of all natural and human-engineered ecosystems, and by extension to the ecosystem services that nature provides free of charge to human society. Living organisms play central roles in the cycles of major elements (carbon, nitrogen, and so on) and water in the environment, and diversity specifically is important in that these cycles require numerous interacting species.

Chapter 9 Biotechnology
9.1 Biotechnology Overview
Exercises

1. Matching

    f d a e b c

2. True or False

    T F T T F

3. Reading Comprehension

    B A B

4. Translation from English to Chinese

　　生物技术利用纯生物科学（遗传学、微生物学、动物细胞培养、分子生物学、生物化学、胚胎学和细胞生物学），并且在很多情况下还依赖于生物学范围以外的知识以及方法（化学工程学、生物加工过程、信息技术和仿生机器人技术）。反之，现代生物科学（甚至包括像分子生态学这样的概念）与通过生物技术及其普遍认为是生命科学领域的知识发展而来的方法是密切相关的，并且依赖于以上的方法。

5. Translation from Chinese to English

　　Biotechnology is not limited to medical/health applications (unlike Biomedical Engineering, which includes much biotechnology). Although not normally thought of as biotechnology, agriculture clearly fits the broad definition of "using a biotechnological system to make products" such that the cultivation of plants may be viewed as the earliest biotechnological enterprise.

9.2　Recombinant DNA Technology

1. Matching

Exercises

　g f e a b c d

2. True or False

　T T F

3. Reading Comprehension

　D B D

4. Translation from English to Chinese

　　一旦载体被大量地分离，它就能被引入哺乳动物、酵母或细菌细胞等宿主细胞。然后宿主细胞将以重组 DNA 为基础合成外源蛋白。当细胞大量生长时，外源蛋白或重组蛋白就可以大量地被分离纯化。

5. Translation from Chinese to English

　　It is important to note that recombinant DNA differs from genetic recombination in that the former results from artificial methods in the test tube, while the latter is a normal biological process that results in the remixing of existing DNA sequences in essentially all organisms. Proteins that result from the expression of recombinant DNA within living cells are termed recombinant proteins.

9.3　Recombinant Protein Expression

Exercises

1. Matching

　f e b d a c

2. True or False

　F T T

3. Reading Comprehension

　A D B

4. Translation from English to Chinese

　　重组蛋白的表达系统主要有两种。一旦我们克隆到 cDNA，就必须决定扩增我们所想要的蛋白质的场所。可以是原核（细菌）表达系统或真核（通常为酵母或哺乳动物细胞）表达系统。由于在大肠杆菌中具有功能的启动子和酵母或哺乳动物细胞具有最佳功能的启动子种类不同，选择的表达系统将决定我们需要用哪个载体对 cDNA 进行克隆。

5. Translation from Chinese to English

　　Prokaryotic recombinant protein expression systems have several advantages. These include ease of culture, and very rapid cell growth meaning you won't have to wait long to get protein from bacterial systems once you clone your cDNA. Expression can be induced easily in bacterial protein expression systems using

IPTG. Also, purification is quite simple in prokaryotic expression systems and there are a plethora of commercial kits available for recombinant protein expression.

9.4 Bioalcohols—Additional Reading

Exercises

1. Matching

   e a d c b

2. True or False

   T F T

3. Reading Comprehension

   A C C

4. Translation from English to Chinese

   甲醇目前是从天然气（一种不可再生的化石燃料）产生的。它也可以从生物量中产生，所产生的甲醇被称为生物甲醇。与当今的天然气制氢技术相比，甲醇经济是通往氢经济的非常令人感兴趣的替代方案。但是这一过程与最新的氢可以直接从水产生的清洁太阳能热能过程并不是一样的。

5. Translation from Chinese to English

   Biologically produced alcohols, most commonly ethanol, and less commonly propanol and butanol, are produced by the action of microorganisms and enzymes through the fermentation of sugars or starches (easiest), or cellulose (which is more difficult). Biobutanol (also called biogasoline) is often claimed to provide a direct replacement for gasoline, because it can be used directly in a gasoline engine (in a similar way to biodiesel in diesel engines).

Chapter 10 Genomics

10.1 The First Sequenced Genomes

Exercises

1. Matching

   b d e a f c

2. True or False

   T T T F T

3. Reading Comprehension

   B B D

4. Translation from English to Chinese

   桑格对φX174 的 DNA 开放阅读框的分析获得了一些意料之外的并且令人吃惊的发现：一些噬菌体基因是重叠在一起的。例如，基因 B 的编码区位于基因 A 内，而基因 E 的编码区位于基因 D 内。此外，基因 D 和基因 J 重叠 1bp。两个基因是如何占据相同位置而编码不同蛋白质的呢？答案是使用不同的密码子进行翻译的两个基因处在这两个阅读框内，这样两个蛋白质产物也就不同。

5. Translation from Chinese to English

   The base sequence of the phage DNA also tells us the amino acid sequence of all the phage proteins. All we have to do is use the genetic code to translate the DNA base sequence of each open reading frame into the corresponding amino acid sequence. This may sound like a laborious process, but a personal computer can do it in a split second.

10.2 The Functional Genomics

Exercises

1. Matching

   b d e a c

2. True or False

F T F

3. Reading Comprehension

C C D

4. Translation from English to Chinese

  功能基因组学包括像突变与多态性（例如 SNP）分析等基因组本身的功能相关的方面，以及分子活性测试的研究。后者由许多诸如转录组学（研究基因表达）、蛋白质组学（研究蛋白质表达）、磷酸化蛋白质组学（蛋白质组学的一个分支）和代谢组学等生物组学组成。功能基因组学主要利用多种技术测试在生物样品中的许多或所有像 mRNA 或蛋白质这样的基因产物的丰度。

5. Translation from Chinese to English

  Unlike genomics and proteomics, functional genomics focuses on the dynamic aspects such as gene transcription, translation, and protein-protein interactions, as opposed to the static aspects of the genomic information such as DNA sequence or structures. Functional genomics attempts to answer questions about the function of DNA at the levels of genes, RNA transcripts, and protein products.

10.3 Proteomics

Exercises

1. Matching

e c d a b

2. True or False

T T F

3. Reading Comprehension

C C A

4. Translation from English to Chinese

  蛋白质组学被认为是继基因组学之后研究生物系统的下一步骤。它比基因组复杂得多，这多是因为生物的基因组或多或少是稳定的，而蛋白质组则因细胞不同而不同，并且随时间变化而变化。这是因为不同基因在不同的细胞类型中进行表达。这意味着即使是一个细胞内产生的基本蛋白质集合也需要被鉴定。

5. Translation from Chinese to English

  Proteomics is the large-scale study of proteins, particularly their structures and functions. Proteins are vital parts of living organisms, as they are the main components of the physiological metabolic pathways of cells. The term "proteomics" was first coined in 1997 to make an analogy with genomics, the study of the genes.

10.4 Bioinformatics—Additional Reading

Exercises

1. Matching

e c d b a

2. True or False

T F T

3. Reading Comprehension

D C D

4. Translation from English to Chinese

  生物信息学的主要目标是增加对生物过程的理解。然而，将其与其他方法区分开的是生物信息学主要是致力于对计算密集型技术的应用和发展（如模型识别、数据挖掘、机器学习算法和可视化）来实现该目标。在这一领域中的主要研究工作包括序列比对、基因发现、基因组拼接、药物设计、药物发现、蛋白质结构比对、蛋白质结构预测、基因表达预测、蛋白质-蛋白质相互作用、基因组范围的协作研究以及进化的建模。

5. Translation from Chinese to English

  Bioinformatics was applied in the creation and maintenance of a database to store biological information at the beginning of the "genomic revolution", such as nucleotide and amino acid sequences. Development of this type of database involved not only design issues but the development of complex interfaces whereby researchers could both access existing data as well as submit new or revised data.

Appendix Ⅱ Index of Professional Words and Phrases

abiotic [eibai'ɔtik] adj. 非生物的
abnormality [æbnɔː'mæliti] n. 异常，畸形
absorptive [əb'sɔːptiv] adj. 吸收性的，有吸收力的
acellular [ei'seljulə] adj. 非细胞的，非细胞组成的
acetogenesis ['æsitəu'dʒenəsis] n. 乙酸形成作用
acetone ['æsətəun] n. 丙酮
acetyl-CoA 乙酰辅酶 A
acid ['æsid] n. 酸，酸性物质；adj. 酸的，酸性的
activation energy 活化能
active site 活性部位
adenine ['ædənin] n. 腺嘌呤
adenosine diphosphate 腺苷二磷酸
adenosine triphosphate 腺苷三磷酸
affliction [ə'flikʃən] n. 痛苦，苦恼，苦难
agarose gel 琼脂糖凝胶
agricultural microbiologist 农业微生物学家
agricultural microbiology 农业微生物学
agricultural productivity 农业生产率
agriculture ['ægrikʌltʃə] n. 农业
AIDS [eidz] abbr. 艾滋病，获得性免疫缺陷综合征（acquired immune deficiency syndrome）
alcohol ['ælkəhɔl] n. 醇类，乙醇
algae ['ældʒiː] n. 水藻，海藻，alga 的复数形式
algal ['ælgəl] adj. 藻类的
algologist [æl'gɔlədʒist] n. 藻类学家
allele [ə'liːl] n. 等位基因
allergy ['ælədʒi] n. 变态反应
allochory ['ælə'tʃɔri] n. 异地传播
amino acid 氨基酸

amino group 氨基
amino terminus 氨基端
ammonia [ə'məunjə] n. 氨
ammonium sulfate 硫酸铵
amphotericin [æmfə'terəsin] n. 两性霉素
amplify ['æmplifai] vt. 扩增
amylase ['æməleis] n. 淀粉酶
anabolism [ə'næbə.lizəm] n. 合成代谢
analog ['ænəlɔːg] n. 类似物，同系物
analogy [ə'nælədʒi] n. 相似，类似
analytical chemistry 分析化学
anatomy [ə'nætəmi] n. 解剖学，解剖
androecium [æn'driʃiəm] n. 雄蕊群
anemochory [ə'niːməkɔːri] n. 风力传播
angiosperm ['ændʒiə'spəːm] n. 被子植物
anhydrase [æn'haidreis] n. 脱水酶
animal husbandry 畜牧业
animalcule ['ænimæl'kjul] n. 微小动物
anther ['ænθə] n. 花粉囊，花药
anthrozoologist [ænθrəzəu'ɔlədʒist] n. 人文动物学者
antibiotic [æntibai'ɔtik] adj. 抗菌的
antibiotics [æntibai'ɔtiks] n. 抗生素
antibody ['æntibɔdi] n. 抗体
anticodon ['ænti'kəudɔn] n. 反密码子
antifungal [ænti'fʌŋgəl] adj. 抗真菌的，杀真菌的；n. 杀真菌剂
antigen ['æntədʒən] n. 抗原
antisense RNA 反义 RNA
aphid ['eifid] n. 蚜虫
apoptotic [æpə'təutik] adj. 细胞凋亡的
applied microbiologist 应用微生物学家
aquatic [ə'kwætik] adj. 水生的

aquatic plant 水生植物
aqueous solution 水溶液
Arabidopsis thaliana 拟南芥
archaea [ɑːˈkiə] n. 古菌
Archaea [ɑːˈkiə] n. 古菌域
archaeological [aːkiəˈlɔdʒikəl] adj. 考古学的
arteriole [ɑːˈtiəriəul] n. 细动脉
artery [ˈɑːtəri] n. 动脉
asexually [eiˈsekʃuəli] adv. 无性地
astrobiology [æstrəˈbaiɔdʒi] n. 天体生物学
ATP synthase ATP 合成酶
autochory [ɔːtəuˈtʃɔri] n. 自动散播
autoimmune [ˌɔːtʊiˈmjuːn] adj. 自身免疫的
autoimmune disease 自身免疫病
Bacillus thuringiensis 苏云金芽孢杆菌
bacteria [bækˈtiriə] n. 细菌，bacterium 的复数形式
Bacteria [bækˈtiriə] n. 细菌域
bacterial [bækˈtiriəl] adj. 细菌的，细菌性的
bacteriologist [bæktiriəˈɔlədʒist] n. 细菌学家
bacteriophage [bækˈtiəriəʃeidʒ] n. 噬菌体
baculovirus cell 杆状病毒细胞
bagasse [bəˈgæs] n. 甘蔗渣
base pairing 碱基配对
basic microbiology 基础微生物学
berry [ˈberi] n. 浆果
beta oxidation β-氧化作用
beta-carotene [beitəˈkærətin] n. β-胡萝卜素
bioalcohol [ˈbaiəuælkəhɔl] n. 生物醇（尤指乙醇）
biobutanol [ˈbaibjuːtənəul] n. 生物丁醇
biochemical [ˈbaiəuˈkemikəl] adj. 生物化学的；n. 生物化工
biochemically [ˈbaiəˈkemikəli] adv. 在生物化学上
biochemistry [ˈbaiəuˈkemistri] n. 生物化学
biocomplexity [ˈbaiəukəmˈpleksəti] n. 生物复杂性
biodiesel [ˈbaiəudiːzl] n. 生物柴油
biodiversity [baiəudaiˈvəːsəti] n. 生物多样性
bioengineering [baiəuendʒəˈniəriŋ] n. 生物工程，生物工程学
bioethanol [ˈbaiəuθənəul] n. 生物乙醇
biofilm [ˈbiɔfilm] n. 生物膜
biofuel [baiəuˈfjuːəl] n. 生物燃料
biogasoline [baiəu ˈgæsəliːn] n. 生化汽油
biogeochemical cycle 生物地球化学循环
biogeography [baiəudʒiˈɔgrəfi] n. 生物地理学
bioinformatics [baiɔinfəˈmætiks] n. 生物信息学
biological control 生物学防治
biological sciences 生物科学
biological systems 生物系统
biological trait 生物性状
biologist [baiˈɔlədʒist] n. 生物学家
biology [baiˈɔlədʒi] n. 生物学
biomass [ˈbaiəuˈmæs] n. 生物量
biomethanol [ˈbaiəuməθənəul] n. 生物甲醇
biomolecule [baiəuˈmɔlikjuːl] n. 生物分子
biopharmaceutical [baiəufɑːməˈsjuːtikəl] n. 生物制药；adj. 生物药剂学的
biophysical [baiəuˈfizikəl] adj. 生物物理的
biophysics [baiəuˈfiziks] n. 生物物理学
bioprocess [baiəuˈprɑːses] n. 生物处理，生物过程
bioprocess engineering 生物加工过程
bioproduct [baiəuˈprɔdəkt] n. 生物制品
bioremediation [baiɔˈrimiːdiˈeiʃən] n. 生物修复
biorobotics [baiəurəuˈbɔtiks] n. 仿生机器人学
biosafety cabinet 生物安全柜
biosphere [ˈbaiəsfiə] n. 生物圈
biosynthetic [baiəusinˈθetik] adj. 生物合成的
biotechnological [baiəutekˈnɔlədʒikəl] adj. 生物技术的
biotechnology [baiəutekˈnɔlədʒi] n. 生物技术
biotic [baiˈɔtik] adj. 生命的，生物的
blood vessel 血管
bond [bɔnd] n. 共价键

botanist ['bɔtənist] n. 植物学家
botany ['bɔtəni] n. 植物学
botulism ['bɔtʃəlizəm] n. 肉毒杆菌中毒
buffer ['bʌfə] 缓冲液
butanol ['bju:tənəul] n. 丁醇
Caenorhabditis elegans 秀丽隐杆线虫
Calvin cycle 卡尔文循环
calyx ['keiliks] n. 花萼
capillary ['kæpiləri] n. 毛细管；adj. 毛细管的
capsule ['kæpsju:l] n. 蒴果, 荚膜
carbohydrate ['kɑ:bəu'haidreit] n. 碳水化合物
carbon ['kɑ:bən] n. 碳
carbon dioxide 二氧化碳
carbon skeleton 碳骨架
carbonic [kɑ:'bɔnik] adj. 碳的
carboxyl terminus 羧基端
cardiovascular disease 心血管疾病
carpel ['kɑ:pəl] n. 心皮, 雌蕊叶
catabolise [kə'tæbəlaiz] v. (使) 分解代谢
catabolism [kə'tæbəlizəm] n. 分解代谢
catalyst ['kætəlist] n. 催化剂
catalyze ['kætəlaiz] v. 催化
causative agent 病原体, 病原物
cDNA 互补 DNA, complementary DNA 的缩写
cell culture 细胞培养
cell cycle arrest 细胞周期阻滞
cell cycle 细胞周期
cell division 细胞分裂
cell junction 细胞连接
cell line 细胞系
cell nucleus 细胞核
cell structure 细胞结构
cell theory 细胞理论
cell wall 细胞壁
cellular ['seljulə] adj. 细胞的
cellular biology 细胞生物学
cellular differentiation 细胞分化
cellular function 细胞功能
cellular senescence 细胞衰老
cellular structure 细胞结构

cellulose ['seljuləus] n. 纤维素
cellulosic [selju'ləusik] adj. 纤维素的
Central Dogma 中心法则
characterize ['kæriktəraiz] v. 鉴定
cheese [tʃi:z] n. 奶酪
chemical bond 化学键
chemical energy 化学能
chemical engineering 化学工程学
chemical reaction 化学反应
chemical structure 化学结构
chemolithotrophy [keməli'θɔtrəfi] n. 化能自养型
chimera [kai'mirə] n. 嵌合体
chlorophyll ['klɔrəfil] n. 叶绿素
chlorophyll ['klɔrəfil] n. 叶绿素
chlorophyll a 叶绿素 a
chloroplast ['klɔ:rəplæst] n. 叶绿体
chromatography [krəumə'tɔgrəfi] n. 色谱法
chromosomal ['krəuməsəuməl] adj. 染色体的
chromosomal analysis 染色体分析
chromosome ['krəuməsəum] n. 染色体
chronic disease 慢性病
circulatory system 循环系统
citric acid cycle 柠檬酸循环, 三羧酸循环
classification [klæsifi'keiʃən] n. 分类, 类别
classification scheme 分类表
classify ['klæsifai] vt. 分类, 归类
Clostridium acetobutylicum 丙酮丁醇梭菌
code [kəud] n. 密码, 法则
coding region 编码区
coding strand 编码链
coenzyme [kəu'enzaim] n. 辅酶
cofactor [kəu'fæktə] n. 辅因子
cold sores 唇疱疹
collagenase ['kɔlədʒəneis] n. 胶原酶
colony ['kɔləni] n. 菌落
community [kə'mju:niti] n. 群落
competent ['kɔmpətənt] adj. 感受态的
complementary [kɔmpli'mentəri] adj. 互补的
computational biology 计算生物学

computational genomics 计算基因组学
connective tissue 结缔组织
consensus sequence 共有序列
construct [ˈkɔnstrʌkt] n. 构建物
contact inhibition 接触抑制
contraceptive [kɔntrəˈseptiv] adj. 避孕的；n. 避孕药，避孕用具
contraction [kənˈtrækʃən] n. 收缩，痉挛
Convention on Biological Diversity 生物多样性公约
copper chloride 氯化铜
corolla [kəˈrɔlə] n. 花冠
cosmid [ˈkɔzmid] n. 黏粒
cotyledon [kɔtəˈliːdən] n. 子叶
Creutzfeldt-Jacob disease 克-雅病
cross-breeding 杂交育种
cross-contamination 交叉污染
cryptozoologist [ˈkriptəuzəuˈɔlədʒist] n. 神秘动物学研究者
cryptozoology [ˈkriptəuzəuˈɔlədʒi] n. 神秘动物学
crystal [ˈkristəl] n. 晶体
crystalline [ˈkristəlain] adj. 晶体的
crystalline lattice 晶格
culture [ˈkʌltʃə] v. 培养；n. 培养，培养物
cyanobacteria [ˈsaiænəuˈbæktiə] n. 蓝细菌，cyanobacterium 的复数形式
cytochrome [ˈsaitəkrəum] n. 细胞色素
cytoplasm [ˈsaitəplæzəm] n. 细胞质
cytosine [ˈsaitəusiːn] n. 胞嘧啶
dairy microbiology 乳品微生物学
dark reaction 暗反应
Darwinian evolution 达尔文进化
deaminate [diˈæməneit] v. （使）脱氨基
deamination [diːæmiˈneiʃən] n. 脱氨基作用
decompose [diːkəmˈpəuz] vi. 分解，腐烂；vt. 腐烂
degradation [degrəˈdeiʃən] n. 退化，降解
dehiscent [diˈhisənt] adj. 裂开性的
dehydration [diːhaiˈdreiʃən] n. 脱水
dehydration reaction 脱水反应

Deinococcus radiodurans 耐辐射球菌
denitrification [diːnaitrifiˈkeiʃən] n. 脱氮作用，反硝化作用
deoxygenate [diːˈɔksədʒəneit] vt. 除去氧气
deoxyribonucleic acid 脱氧核糖核酸
deoxyribose [diːˌɔksiˈraibəus] n. 脱氧核糖
derivative [diˈrivətiv] n. 衍生物
development [diˈveləpmənt] n. 发育
developmental biology 发育生物学
dicotyledon [daikɔtəˈliːdən] n. 双子叶植物
dideoxy chain termination method 双脱氧链终止法
differentiation [difərenʃiˈeiʃən] n. 分化
digestion [diˈdʒestʃən] n. 消化，吸收
digestive system 消化系统
digestive tract 消化道
dilute [daiˈluːt] vt. 稀释，冲淡
disaccharide [daiˈsækəraid] n. 二糖
disease [diˈziːz] n. 疾病
disease resistance 抗病性
disease-resistant adj. 抗病的
distillation [distəˈleiʃən] n. 蒸馏
diversity [daiˈvəːsiti] n. 多样性，差异
divide [diˈvaid] vi. 分裂，分开
division [diˈviʒən] n. 分裂，分开
DNA analysis （法医的）DNA 分析
DNA fingerprinting DNA 指纹图谱
DNA insert DNA 插入
DNA ligase DNA 连接酶
DNA polymerase DNA 聚合酶
DNA replication DNA 复制
DNA sequencing DNA 测序
domestic [dəˈmestik] adj. 驯养的
dose [dəus] n. 剂量
double helix structure 双螺旋结构
Down's syndrome 唐氏综合征
dread [dred] adj. 可怕的，恐怖的；v. 恐惧，害怕
drug resistance 药物抗性
drug-screening 药物筛选
drupe [druːp] n. 核果

duplex RNA   双链 RNA
dynamics   [dai'næmiks]   n. 力学，动力学，动态
echinoderm   [e'kainədə:m]   n. 棘皮动物类的动物
ecological dynamics   生态动力学
ecologist   [i:'kɔlədʒist]   n. 生态学家
ecology   [i:'kɔlədʒi]   n. 生态学
ecosystem   ['i:kəusistəm]   n. 生态系统
ecotone   ['i:kətəun]   n. 群落交错区
ectoderm   ['ektəudə:m]   n. 外胚层
electromagnetic radiation   电磁辐射
electron acceptor   电子受体
electron microscope   电子显微镜
electron transport chain   电子传递链
electroporation   [i'lektrəupɔreiʃən]   n. 电转化
elephantiasis   [eləfən'taiəsis]   n. 象皮病
embryo   ['embriəu]   n. 胚胎
embryology   [embri'ɔlədʒi]   n. 胚胎学，发生学
embryonic   [embri'ɔnik]   adj. 胚胎的
emphysema   [emfi'si:mə]   n. 肺气肿
encode   [in'kəud]   vt. 编码
end product   终产物
endocarp   ['endəukɑ:p]   n. 内果皮
endocrine   ['endəukrain]   adj. 内分泌的
endocrine system   内分泌系统
endoderm   ['endəudə:m]   n. 内胚层
endosperm   ['endəspə:m]   n. 胚乳
enhancer   [in'hɑ:nsə]   n. 增强子
entropic effect   熵效应
environmental control   环境控制
environmental pollution   环境污染
environmental science   环境科学
enzymatic digestion   酶消化
enzyme   ['enzaim]   n. 酶
enzyme activity   酶活性
Ephedra vulgaris   麻黄
epidermal cell   表皮细胞
epidermis   [,epi'də:mis]   n. 表皮，上皮
epithelial   [epi'θi:liəl]   adj. 上皮的

equilibrium   [i:kwi'libriəm]   n. 平衡
erosion   [i'rəuʒən]   n. 腐蚀，侵蚀
Escherichia coli   大肠杆菌
estrogen   ['estrədʒən]   n. 雌激素
ethanol   ['eθənəul]   n. 乙醇
ethology   [i:'θɔlədʒi]   n. 动物行为学
eugenics   [ju:'dʒeniks]   n. 优生学
Eukarya   [ju:'kæriə]   n. 真核生物域
eukaryote   [ju'kæriəut]   n. 真核生物
eukaryotic   [ju'kæriəutik]   adj. 真核细胞的
eukaryotic kingdom   真核生物界
eukaryotic microbes   真核微生物
Eumycota   ['jumaikəutə]   n. 真菌门
evolution   [i:və'lu:ʃən]   n. 进化
evolutionary   [evə'lu:ʃən.eri]   adj. 进化的
evolutionary biology   进化生物学
evolutionary tree   进化树
ex vivo   （与培养有关的）体外
excretion   [eks'kri:ʃən]   n. 排泄
exocarp   ['eksəukɑ:p]   n. 外果皮
exocrine   ['eksəkrin]   adj. 外分泌的
experimentation   [iksperimen'teiʃn]   n. 实验方法
express   [iks'pres]   vt.（基因）使表达
expression   [iks'preʃən]   n.（基因）表达
extracellular   [ekstrə'seljulə]   adj. (位于或发生于)细胞外的
extreme   [iks'tri:m]   adj. 极度的, 极端的; n. 极端, 极限
extreme environment   极端环境
farm productivity   农业生产力
fatty acid   脂肪酸
feedback   ['fi:dbæk]   n. 反馈
feedstock   ['fi:dstɔk]   n. 原料
ferment   ['fə:ment]   v. 发酵
fermentation   [fə:men'teiʃən]   n. 发酵
ferrous   ['ferəs]   n. 亚铁，二价铁
fertilization   [fə:tilai'zeiʃən]   n. 受精, 肥沃化
fiber   ['faibə]   n. 纤维
filament   ['filəmənt]   n. 花丝

filtration [fil'treiʃən] n. 过滤
fishery ['fiʃəri] n. 渔场，渔业
five-carbon sugar 五碳糖
follicle ['fɔlikl] n. 骨突果，卵泡细胞
food science 食品科学
food-borne ['fu:d bɔ:n] adj. 食物传染的
foreign DNA 外源 DNA
foreign gene 外源基因
forestry ['fɔristri] n. 林业
fossilize ['fɔsilaiz] vt. & vi. 使成化石，使陈腐
fructose ['frʌktəus] n. 果糖
functional ['fʌŋkʃənəl] adj. 功能的，实用的
functional genomics 功能基因组学
fungal toxin 真菌毒素
fungi ['fʌndʒai] n. 真菌，fungus 的复数形式
fungicide ['fʌndʒisaid] n. 杀真菌剂
gametangia ['gæmə'tændʒiəm] n. 配子囊
gamete ['gæmi:t] n. 配子
gap junction 缝隙连接
gel electrophoresis 凝胶电泳
gene cloning 基因克隆
gene expression 基因表达
gene product 基因产物
gene regulation 基因调控
gene sequence 基因序列
gene structure 基因结构
gene theory 基因理论
gene transfer 基因转移
genetic cloning 遗传（基因）克隆
genetic code 遗传密码
genetic disease 遗传疾病
genetic engineering 遗传工程
genetic information 遗传信息
genetic makeup 基因组合
genetic manipulation 遗传（基因）操作
genetic material 遗传物质
genetic program 遗传程序
genetic recombination 遗传重组
genetic system 遗传系统
genetic trait 遗传性状

genetics [dʒi'netiks] n. 遗传学
genome ['dʒi:nəum] n. 基因组
genome assembly 基因组拼接
genomic mapping 基因组作图
genomics [dʒə'nəumiks] n. 基因组学
genus ['dʒi:nəs] n. （分类学上的）属
geological [dʒiə'lɔdʒikəl] adj. 地质学的
geological condition 地质条件
germ [dʒə:m] n. 病原微生物
germ cell 生殖细胞，受精卵
germ line 种系
germination [dʒə:mi'neiʃən] n. 发芽，萌芽
geyser ['gaizə] n. 间歇喷泉
gland [glænd] n. 腺
glucogenic [glu:kə'dʒenik] adj. 生成葡糖的
gluconeogenesis [glu:kəuni:ə'dʒenəsis] n. 糖异生作用
glucose ['glu:kəus] n. 葡萄糖
glutamate ['glu:təmeit] n. 谷氨酸盐
glyceraldehyde 3-phosphate 3-磷酸甘油醛
glycerol ['glisərəul] n. 甘油
glycolysis [glai'kɔləsis] n. 糖酵解
glycoproteomics ['glaikəuprəuti:əumiks] n. 糖蛋白质组学
gonad ['gɔnæd] n. 性腺
grain [grein] n. 谷物，谷类
grana ['greinə] n. 基粒
granulosa [grænju'ləusə] n. 粒层细胞
green fluorescence protein 绿色荧光蛋白
green plants 绿色植物
greenhouse ['gri:nhaus] n. 温室
growth factor 生长因子
guanine ['gwɑ:ni:n] n. 鸟嘌呤
gymnosperm ['dʒimnəspə:m] n. 裸子植物
gynoecium [dʒi'ni:siəm] n. 雌蕊群
habitat ['hæbitæt] n. 栖息地，产地
habitat fragmentation 生境破碎化
haemophilia [hi:mə'filiə] n. 血友病，出血不止
Haemophilus influenzae 流感嗜血杆菌
hanging drop method 悬滴法

hantavirus ['hæntə'vairəs] n. 汉坦病毒
Hayflick limit 海弗利克极限(指培养中细胞生命的自然极限)
heat-tolerant adj. 耐热的
hemorrhagic ['heməræd3ik] adj. 出血的
hemorrhagic fever 出血热
herbaceous [hə:'beiʃəs] adj. 草本植物的
herbicide ['hə:bəsaid] n. 除草剂
herbicide resistance 除草剂抗性
hereditary [hi'reditəri] adj. 世袭的, 遗传的
heredity [hi'rediti] n. 遗传
hierarchical ['haiə'rɑ:kikəl] adj. 按等级划分的
high-throughput method 高通量方法
His tag 组氨酸标签
histone ['histəun] n. 组蛋白
homeostasis [həumiə'steisis] n. 自动平衡, 体内平衡
horizontal gene transfer 水平基因转移
hormone ['hɔ:məun] n. 荷尔蒙, 激素
host [həust] n. 宿主
host cell 宿主细胞
human ecology 人类生态学
human lymphocyte antigen typing 人淋巴细胞抗原分型
hybridization [haibridai'zeiʃən] n. 杂交
hydrochory ['haidrəkɔ:ri] n. 水流传播
hydrogen ['haidrədʒən] n. 氢
hydrogen bond 氢键
hydrogen sulfide 硫化氢
hydrologic system 水文系统
hydroxyl group 羟基
hypothesis [hai'pɔθəsis] n. 假说, 假设, 猜测
ichthyologist [ikθi'ɔlədʒist] n. 鱼类学者
immune system 免疫系统
immunology [imju'nɔlədʒi] n. 免疫学
impervious [im'pə:viəs] adj. 不能渗透的, 不为所动的
in vitro 在试管内, 体外
in vivo 体内
inanimate world 非生命界

inclusion body 包涵体
incubator ['inkjubeitə] n. 培养箱
indication [indi'keiʃən] n. 指示, 表示, 迹象
individual [indi'vidjuəl] n. 个体
induced fit hypothesis 诱导契合假说
industrial microbiology 工业微生物学
inequality [ini'kwɔləti] n. 不平等, 不平均, 不平坦
infection [in'fekʃən] n. 传染, 侵染
infectious agents 传染源, 传染物
infectious disease 传染病
inflorescence [infləu'resəns] n. 花序
information technology 信息技术
inheritance [in'heritəns] n. 遗传, 继承物
initiation codon 起始密码子
inorganic [inɔ:'gænik] adj. 无机的, 无生物的
inorganic compound 无机化合物
insect ['insekt] n. 昆虫, 虫子
insect resistance 抗虫性
instrumentation [instrəmen'teiʃən] n. 仪器, 仪表
insulin ['insjulin] n. 胰岛素
integument [in'tegjumənt] n. 外皮
integumentary system 皮肤系统
interaction [intə'rækʃən] n. 相互关系, 相互作用
intermediate [intə'mi:diət] n. 中间产物
intermembrane space 膜间腔
interrelation [intəri'leiʃən] n. 相互关系
intracellular [intrə'seljulə] adj. 细胞内的
intron ['intrɔn] n. 内含子
invasion [in'veiʒən] n. 侵略, 侵入
IPTG 异丙基硫代-β-D-半乳糖苷 (isopropyl-beta-D-thiogalactopyranoside)
isoenzyme [aisəu'enzaim] n. 同工酶
isoenzyme analysis 同工酶分析
isolate ['aisəleit] vt. 使分离; n. 分离物
isotope ['aisətəup] n. 同位素
IUCN Red List 世界自然保护联盟（International Union for Conservation of Nature）红色名录
junk DNA 垃圾 DNA

jute [dʒuːt] n. 黄麻
karyotyping ['kæriətaipiŋ] n. 染色体组型分型
keto acid 酮酸
α-ketoglutarate ['ælfə kiːtəu'gluːtəreit] α-酮戊二酸
kidney ['kidni] n. 肾
kingdom ['kiŋdəm] n. （分类学上的）界
kingdom Fungi 真菌界
kit [kit] n. 试剂盒
knock out 敲除（基因）
knockout ['nɔkaut] n. （基因）敲除
laboratory ['læbrətɔːri] n. 实验室
lactate ['lækteit] n. 乳酸
lactate dehydrogenase 乳酸脱氢酶
lactic acid 乳酸
laminar flow hood 层流净化罩
lichen ['laikin] n. 地衣，青苔
lichenology [laikə'nɔlədʒi] n. 地衣学
life science 生命科学
light reaction 光反应
lipid ['lipid] n. 脂
livestock ['laivstɔk] n. 家畜，牲畜
lutein ['luːtiin] n. 黄体素，叶黄素
lymph [limf] n. 淋巴，淋巴液
lymph vessel 淋巴管
lymphatic system 淋巴系统
lymphocyte ['limfəsait] n. 淋巴细胞
lymphoma [lim'fəumə] n. 淋巴癌
lyse [lais] v. 裂解（细胞）
lysozyme ['laisəzaim] n. 溶菌酶
macromolecule [mækrə'mɔləkjuːl] n. 大分子
magnesium ['mæg'niːziəm] n. 镁
mammalian [mæ'meiljən] adj. 哺乳动物的
mammalian cell 哺乳动物细胞
mammalogist [mæ'mælədʒist] n. 哺乳动物学者
Mars [mɑːz] n. 火星
matrix ['meitriks] n. 基质
mature transcript 成熟转录本
mechanical philosophy 机械论哲学
mechanism ['mekənizəm] n. 机制，原理

media ['miːdiə] n. 培养基，medium 的复数形式
medical ['medikəl] adj. 医疗的，医学的，医药的，内科的
medical advancement 医学进展
medical microbiologist 医学微生物学家
medical microbiology 医学微生物学
medicinal [mə'disənəl] adj. 药物的
medicine ['medisin] n. 医学
megasporophyll [megə'spɔːrəfil] n. 大孢子叶
membrane ['membrein] n. 薄膜，膜状物
meristem ['meristem] n. 分生组织
mesocarp ['mesəkɑːp] n. 中果皮
mesoderm ['mesədəːm] n. 中胚层
mesophyll ['mesəfil] n. 叶肉
mesophyll cell 叶肉细胞
messenger RNA 信使 RNA
metabolic activity 代谢活动
metabolic pathway 代谢途径
metabolic process 代谢过程
metabolism [me'tæbəlizəm] n. 新陈代谢
metabolomics [metæ'bɔləmiks] n. 代谢组学
methane ['meθein] n. 甲烷，沼气
Methanococcus janaschii 詹式甲烷球菌
methanol ['meθənəul] n. 甲醇
microbe ['maikrəub] n. 微生物
microbe-microbe interaction 微生物与微生物相互作用
microbial community 微生物群落
microbial cytology 微生物细胞学
microbial diversity 微生物多样性
microbial ecologist 微生物生态学家
microbial ecology 微生物生态学
microbial energy 微生物能源
microbial geneticist 微生物遗传学家
microbial genetics 微生物遗传学
microbial growth 微生物生长
microbial morphology 微生物形态学
microbial physiology 微生物生理学
microbial taxonomy 微生物分类学
microbiologist [maikrəubai'ɔləgist] n. 微生物

学家
microbiology [maikrəubai'ɔlədʒi] n. 微生物学
microcarrier [maikəu'kæriə] n. 微载体
microinjection [maikrəuin'dʒekʃən] n. 显微注射
microorganism [maikrəu'ɔ:gənizəm] n. 微生物
microscope ['maikrəskəup] n. 显微镜
microscopic ['maikrə'skɔpik] adj. 显微镜的, 微观的
microscopist [mai'krɔskəpist] n. 使用显微镜的技术人员
microscopy ['maikrəskəupi] n. 显微镜使用, 用显微镜检查
microsphere ['maikrəsfiə] n. 微球体, 微滴
microsporophyll [maikrə'spɔ:rəfil] n. 小孢子叶
mitochondria [maitə'kɔndriə] n. 线粒体, mitochondrium 的复数形式
modern microbiology 现代微生物学
molasses [mə'læsiz] n. 糖浆, 糖蜜
mold [məuld] n. 霉菌
molecular biochemistry 分子生物化学
molecular biology 分子生物学
molecular ecology 分子生态学
molecular level 分子水平
molecular sequence 分子序列
molecular technique 分子技术
molecule ['mɔlikju:l] n. 分子
Monera [mə'niərə] n. 无核原虫界
monoclonal antibody 单克隆抗体
monocotyledon [mɔnəkɔtəl'i:dən] n. 单子叶植物
monomer ['mɔnəmə] n. 单体
mononuclear ['mɔnəu'nju:kliə] adj. 单核的
monophyletic [mɔnəufai'letik] adj. 单元的, 单源的
monosaccharide [mɔnə'sækəraid] n. 单糖
morphological [mɔ:fə'lɔdʒikəl] adj. 形态学的
mRNA analysis mRNA 分析
multicellular [mʌlti'seljulə] adj. 多细胞的
muscular system 肌肉系统

mutability [mju:tə'biləti] n. 易变性
mutation [mju:'teiʃən] n. 突变
Myc tag Myc （一个癌基因）标签
mycologist [mai'kɔlədʒist] n. 真菌学家
mycology [mai'kɔlədʒi] n. 真菌学
Mycoplasma genitalium 生殖支原体
mycoses [mai'kəusis] n. 霉菌病, mycosis 的复数形式
mycotoxicology [maikəutɔksə'kɔlədʒi] n. 真菌毒理学
myxobacteria [miksəbæk'tiəriə] n. 黏细菌
natural selection 自然选择
natural variation 自然变异
naturalist ['nætʃərəlist] n. 自然主义者
nature conservation 自然保护
necrotic [ne'krɔtik] adj. 坏死的, 坏疽的
nerve [nə:v] n. 神经
nervous system 神经系统
nested hierarchy 包含型等级系统
niche [nitʃ] n. 生境
nitrification [naitrəfi'keiʃən] n. 硝化作用
nitrite ['naitrait] n. 亚硝酸盐
nitrogen ['naitrədʒən] n. 氮
nitrogen fixation 固氮作用
noncellular [nɔn'seljulə] adj. 非细胞的
noncovalent bond 非共价键
noncovalent interaction 非供价作用
nondehiscent [nʌndi'hisənt] adj. 不裂开性的
nonfertile ['nʌn'fə:tail] adj. 不育的, 不孕的
nongenic element 非基因元件
novel ['nɔvəl] adj. 新奇的
nucellus [nu:'seləs] n. 珠心
nucleic acid 核酸
nucleophile ['nju:kli:əfail] n. 亲核试剂
nucleotide ['nju:kliətaid] n. 核苷酸
nucleotide sequence 核苷酸序列
nucleus ['nju:kliəs] n. 细胞核
nutrient source 营养源
nutriment ['nju:trimənt] n. 营养品, 养料, 滋养物

octane ['ɔktein] n. 辛烷
offspring ['ɔ:fspriŋ] n. 后代，子孙
omics ['əumiks] n. 生物组学
Onychophora [ɔni'kɔfərə] n. 有爪动物门
oocyte ['əuəsait] n. 卵母细胞
open reading frame 开放阅读框
operon ['ɔpə,rɔn] n. 操纵子
optics ['ɔptiks] n. 光学
organ ['ɔ:gən] n. （生物的）器官
organ system 器官系统
organelle [ɔ:gə'nel] n. 细胞器
organic [ɔ:'gænik] adj. 器官的，有机的
organic compound 有机化合物
organism ['ɔ:gənizəm] n. 生物体
organismal biology 有机体生物学
organotypic ['ɔ:gənəu'tipik] adj. 器官型的
organotypic culture 器官型培养
osmotic [ɔz'mɔtik] adj. 渗透的
ovary ['əuvəri] n. 卵巢，子房
overlap ['əuvə'læp] n. 重叠，重复
oxidative phosphorylation 氧化磷酸化
oxidizer ['ɔksədaizə] n. 氧化剂
oxygenate ['ɔksidʒineit] v. 以氧处理，氧化；n. 氧化剂
paleobotanist [peiliə'bɔtənist] n. 古植物学家
paleontologist [,pæliɔn'tɔlədʒist] n. 古生物学家
paleozoologist [peiliəuzəu'ɔlədʒist] n. 古动物学家
paleozoology [peiliəuzəu'ɔlədʒi] n. 古动物学
parasite ['pærəsait] n. 寄生虫，寄生生物
parasitic [pærə'sitik] adj. 寄生的
parent ['pəərənt] n. 亲本，母本
passage ['pæsidʒ] n. & v. 传代
pathogen ['pæθədʒen] n. 病原体
pathogenic [pæθə'dʒenik] adj. 致病的
pathologist [pæ'θɔlədʒist] n. 病理学家
pedicel ['pedəsəl] n. 花梗，肉茎
pedogenesis [pi:dəu'dʒenəsis] n. 土壤发生
peduncle [pi'dʌŋkl] n. 花梗，肉茎
penicillin [peni'silin] n. 青霉素

Penicillium [peni'siliəm] n. 青霉菌属
pentose ['pentəus] n. 戊糖
peptidase ['peptideis] n. 肽酶
peptide ['peptaid] n. 肽
peptide bond 肽键
perianth ['periænθ] n. 花被
pericarp ['perikɑ:p] n. 果皮
perisperm ['perispə:m] n. 外胚乳
pest control 昆虫防治
pesticide ['pestisaid] n. 杀虫剂，农药
petal ['petl] n. 花瓣
pH indicator pH 指示剂
phage [feidʒ] n. 噬菌体
phallus ['fæləs] n. 阴茎
pharmaceutical [fɑ:mə'sju:tikəl] adj. 制药的, n. 药品
phenotype ['fi:nətaip] n. 表型
phenotypic ['fi:nətipik] adj. 表型的
phosphate ['fɔsfeit] n. 磷酸盐
phosphoproteomics ['fɔsfəuprəuti:əumiks] n. 磷酸化转录组学
phosphorylate ['fɔsfərileit] vt. 使磷酸化
photosynthesis [fəutəu'sinθəsis] n. 光合作用
photosynthetic [fəutəusin'θetik] adj. 光合的
photosynthetic pigment 光合色素
phycologist [fai'kɔlədʒist] n. 藻类学家
phyla ['failə] n. （分类学上的）门，phylum 的复数形式
phylogenetic [failəudʒə'netik] adj. 系统发生的
phylogenetic tree 系统发育树
physical anthropology 体质人类学
physiological [fiziə'lɔdʒikəl] adj. 生理的, 生理学的
physiologically [fiziə'lɔdʒikəli] adv. 在生理学上
physiology [fizi'ɔlədʒi] n. 生理学
pickle ['pikəl] n. 泡菜
pigment ['pigmənt] n. 色素
pistil ['pistil] n. 雌蕊
planetary ['plænitəri] adj. 行星的
plasmid ['plæzmid] n. 质粒

plating density　接种密度
pod　[pɔd]　n. 荚果
pollen　['pɔlən]　n. 花粉
pollinate　['pɔlineit]　vt. 给……授粉
pollutant　[pə'lu:tənt]　n. 污染物
polyethylene glycol　聚乙二醇
polymer　['pɔlimə]　n. 多聚体
polymerase　['pɔliməreis]　n. 聚合酶
polymerase chain reaction　聚合酶链式反应
polymorphism　[pɔli'mɔ:fizəm]　n. 多态性
polynucleotide　[pɔli'nju:kljətaid]　n. 多聚核苷酸
polynucleotide kinase　多核苷酸激酶
polypeptide chain　多肽链
polysaccharide　[pɔli'sækə.raid]　n. 多糖
population　[pɔpju'leiʃən]　n. 种群
population genetics　群体遗传学
porphyrin　['pɔ:fərin]　n. 卟啉
post-transcriptional　['pəust træn'skripʃənəl] adj. 转录后的
post-translational modification　翻译后修饰
potassium　[pə'tæsiəm]　n. 钾
precipitant　[pri'sipitənt]　n. 沉淀剂
precipitation　[prisipi'teiʃən]　n. 沉淀
primary transcript　初级转录本
primer　['praimə]　n. 引物
primitive　['primitiv]　adj. 原始的, 简陋的
primordia　[prai'mɔ:djə]　n. 原基, 原始细胞, primordium 的复数形式
progesterone　[prəu'dʒestərəun]　n. 黄体酮
prokaryote　[prəu'kæriɔt]　n. 原核生物
prokaryotic　[prəukæri'ɔtik]　adj. 原核的
promoter　[prə'məutə]　n. 启动子
pronase　['prəuneis]　n. 链霉蛋白酶
propanol　['prəupənɔl]　n. 丙醇
protein　['prəuti:n]　n. 蛋白质
protein content　蛋白质含量
protein crystal　蛋白质晶体
protein degradation rate　蛋白质降解率
protein domain　蛋白质结构域
protein sequence　蛋白质序列

protein structure alignment　蛋白质结构比对
protein structure prediction　蛋白质结构预测
protein structure　蛋白质结构
protein-protein interaction　蛋白质与蛋白质之间的相互作用
proteome　['prəuti:əum]　n. 蛋白质组
proteomics　['prəuti:əumiks]　n. 蛋白质组学
protist　['prəutist]　n. 原生生物
protista　[prəu'tistə]　n. 原生动物界
protistan　[prəu'tistən]　adj. 原生生物的
protocell　n. 细胞的原始状态
proto-genome　n. 原始的基因组
protozoa　[prəutə'zəuə]　n. 原生动物, protozoan 的复数形式
protozoologist　[prəutəzəu'ɔlədʒist]　n. 原生动物学家
protozoology　[prəutəzəu'ɔlədʒi]　n. 原生动物学
Pseudomonas　[(p)sju:'dɔmənəs]　n. 假单胞菌属
public health microbiologist　公共健康微生物学家
public health microbiology　公共健康微生物学
pulse　[pʌls]　n. 豆类等结荚植物的可食性种子
pyruvate　[pai'ru:veit]　n. 丙酮酸盐
quasi-stable state　准稳态
quintillion　[kwin'tiljən]　n. 百万的三次方
rachis　['reikis]　n. 花轴
random mutation　随机突变
reactant　[ri'æktənt]　n. 反应物
receptor　[ri'septə]　n. 受体
recombinant DNA　重组 DNA
recombinant DNA technology　重组 DNA 技术
recombinant protein　重组蛋白
recombinase　[ri:'kɔmbə'neis]　n. 重组酶
reducing power　还原力
regulatory mechanism　调控机制
regulatory region　调控区
replication　[repli'keiʃən]　n. 复制
reporter protein　报告蛋白
repository　[ri'pɔzətəuri]　n. 容器, 仓库, 贮藏室
repressor　[ri'presə]　n. 阻遏物
reproduce　[ri:prə'dju:s]　v. 再生, 复制, 生殖

reproductive system 生殖系统
respiration [respə'reiʃən] n. 呼吸
respiratory syncytial virus 呼吸道合胞病毒
respiratory system 呼吸系统
restriction endonuclease 限制性内切酶
restriction enzyme 限制性酶
retrovirus [retrəu'vaiərəs] n. 逆转录病毒
rheumatoid arthritis 类风湿性关节炎
rhodopsin [rəu'dɔpsin] n. 视紫质
ribonuclease [raibəu'nju:klieis] n. 核糖核酸酶
ribose ['raibəus] n. 核糖
ribosome ['raibəsəum] n. 核糖体
RNA polymerase RNA 聚合酶
roundworm ['raundwə:m] n. 蛔虫
rudimentary [,ru:də'mentəri:] adj. 基本的，初步的，未充分发展的
ruminant ['ru:minənt] n. 反刍动物
Saccharomyces cerevrsiae 酿酒酵母
salmonellosis [sælmənə'ləusis] n. 沙门氏菌病
salt tolerance 耐盐性
sample ['sæmpl] n. 样品，样本；vt. 采样，取样
secretion [si'kri:ʃən] n. 分泌
sediment ['sedimənt] n. 沉淀
self-replicate ['self'replikeit] v. 自我复制
self-replication n. 自我复制
senescence [sə'nesəns] n. 衰老
sensor ['sensə] n. 传感器
sepal ['si:pəl] n. 萼片
sequence alignment 序列比对
serum ['sirəm] n. 血清
sexual reproduction 有性生殖
sexually ['sekʃuəli] adv. 有性地
shoots [ʃu:ts] n. 嫩枝
short tandem repeat (STR) 短串联重复
signal transduction 信号转导
sitting drop method 静滴法
sodium/proton antiport pump 钠/氢逆向转运泵
soft tissue 软组织
solar energy 太阳能
solvent ['sɔlvənt] n. 溶剂

specialty ['speʃəlti] n. 专业，专长
species ['spi:ʃi:z] n. 种类，（单复同）物种
sperm [spə:m] n. 精子
spleen [spli:n] n. 脾
splice [splais] vt. 剪切，拼接
splicing ['splaisiŋ] n. 剪切，拼接
splitting cells 分细胞
spontaneous generation 自然发生
spore-bearing adj. 产生孢子的
stalk [stɔ:k] n. 茎，梗
stamen ['steimən] n. 雄蕊
starch [stɑ:tʃ] n. 淀粉
stem [stem] n. 茎
stem cell 干细胞
sterile technique 无菌操作技术
sterilize ['sterilaiz] vt. 使不育，杀菌
steroid ['stiərɔid] n. 类固醇
stigma ['stigmə] n. 柱头
stomate ['stəumeit] n. 气孔
stop codon 终止密码子
strain [strein] n. 菌株
strata ['streitə] n. 地层
streptomycin ['streptə'maisin] n. 链霉素
stress [stres] n. 压力，胁迫
stroma ['strəumə] n. 基质，子座
stromalite [strəu'mætlait] n. 叠层石
strong promoter 强启动子
subcellular [sʌb'seljulə] adj. 亚细胞的
subculture ['sʌbkʌltʃə] n. 继代培养
submission [sʌb'miʃən] n. 投（稿），提交
substrate ['sʌbstreit] n. 底物
subunit [sʌb'ju:nit] n. 亚基，亚单位
sucrose ['sju:krəus] n. 蔗糖
sugar beet 糖用甜菜
sugar cane 甘蔗
sulfide ['sʌlfaid] n. 硫化物
sulfur ['sʌlfə] n. 硫
sulfur compound 含硫化合物
supracrustal [sju:prə'krʌstəl] adj. (地层、岩组等)覆盖基底岩石的，上地壳的

survival [sə'vaivəl] n. 生存，存活
symbiotic relationship 共生作用
synthase ['sinθeis] n. 合成酶
synthetic biology 合成生物学
synthetic DNA 合成 DNA
taxa ['tæksə] n. 分类单元，taxon 的复数形式
taxonomic [tæksə'nɔmik] adj. 分类学的
taxonomy [tæk'sɔnəmi] n. 分类学
telomere ['teləmiə] n. 端粒
template strand 模板链
terminator gene 终止子基因
terrestrial [ti'restriəl] adj. 陆地的，陆生的
testa ['testə] n. 外种皮
the origin of life 生命的起源
the Royal Society （英国）皇家学会
theca ['θi:kə] n. 泡膜
theology [θi'ɔlədʒi] n. 神学
theoretical biology 理论生物学
thermal efficiency 热效率
Thermus aquaticus 水生栖热菌
thiosulfate [θaiəu'sʌlfeit] n. 硫代硫酸盐
thorium ['θɔ:riəm] n. 钍
three-carbon compound 三碳化合物
thylakoid ['θailəkɔid] n. 类囊体
thymine ['θaimi:n] n. 胸腺嘧啶
tissue ['tiʃu:] n. （生物的）组织
tissue culture 组织培养
toxic waste 有毒废物
toxin ['tɔksin] n. 毒素
trait [treit] n. 性状，品质
transaminase [træn'sæmineis] n. 转氨酶
transcribe [træns'kraib] v. 转录
transcript ['trænskript] n. 转录本，转录物
transcription [træns'kripʃən] n. 转录
transcription initiation site 转录起始位点
transcription machinery 转录复合物
transcriptomics ['trænskriptəumiks] n. 转录组学
transduction [træns'dʌkʃən] n. 转导
transfect [træns'fekt] vt. 转染

transfer RNA 转运 RNA
transformation [trænsfə'meiʃən] n. 转化
transgene [trænz'dʒi:n] n. 输入基因
transgenic [,trænz'dʒenik] n. 转基因（做法），转基因学；adj. 转基因的
transgenic animal 转基因动物
translate [træns'leit] v. （RNA）翻译
translation [træns'leiʃən] n. （RNA）翻译
translation machinery 翻译机
trypsin ['tripsin] n. 胰蛋白酶
ultrastructural ['ʌltrəl'strʌktʃərəl] adj. 超微结构的
unicellular ['ju:ni'seljulə] adj. 单细胞的
unidentified ['ʌnai'dentifaid] adj. 未确认的，无法识别的
uracil ['jurəsil] n. 尿嘧啶
uranium [juə'reiniəm] n. 铀
urban ecology 城市生态学
urea [ju'ri:ə] n. 尿素
urea cycle 尿素循环
urinary ['jurəneri] adj. 泌尿（器官）的，尿的
urinary system 泌尿系统
uterus ['ju:tərəs] n. 子宫
vaccine ['væksi:n] n. 疫苗
vacuole ['vækjuəul] n. 液泡
vapor diffusion 蒸汽扩散
variability [veriə'biləti] n. 变化性，变异性
variant ['veriənt] n. 变体，变型
variation [veri'eiʃən] n. 变种，变化
vascular ['væskjulə] adj. 脉管的
vascular bundle 维管束
vascular conducting system 维管系统
vector ['vektə] n. 载体
vegetation association 植物群丛
vein [vein] n. 静脉，叶脉
velvet worm 栉蚕
vernalize ['və:nəlaiz] v. 施以春化处理，以人工方法促进发育
vertebrate ['və:tibreit] n. 脊椎动物
viability [vaiə'biliti] n. 生存能力，发育能力

viral ['vairəl] adj. 病毒的
virion ['vaiərion] n. 病毒粒子
virologist [vai'rɔləgist] n. 病毒学家
virus ['vairəs] n. 病毒，病原体
virus particle 病毒粒子
vitamin ['vaitəmin] n. 维生素
weak promoter 弱启动子
well-being ['wel'bi:iŋ] n. 健康，福利
whorl [wə:l] n. 轮生体

Wuchereria bancrofti 班克罗夫特线虫
X-ray crystallography X 射线晶体衍射学
X-ray diffraction X 射线衍射
yeast [ji:st] n. 酵母，发酵剂
Yellowstone National Park （美国）黄石国家公园
yogurt ['jəugə:t] n. 酸奶
zoochory ['zəuəkəri] n. 动物传播
zoologist [zəu'ɔlədʒist] n. 动物学家
zoology [zəu'ɔlədʒi] n. 动物学

References

[1] Enger E D, Ross F C, Bailey D B. Concepts in Biology. Thirteenth Edition. Boston: McGraw Hill Higher Education, 2009.

[2] Campbell N A, Reece J B, Simon E J. Essential Biology. Third Edition. San Francisco: Pearson Education, Inc, 2007.

[3] Prescott L M, Harley J D, Klein D A. Microbiology. Seventh Edition. Boston: McGraw Hill Higher Education, 2008.

[4] Karp G. Cell and Molecular Biology: Concepts and Experiments. Fifth Edition. New York : John Wiley &Sons Inc, 2007.

[5] Stern K R, Jansky S H, Bildack J E. Introductory Plant Biology. Twelfth Edition. Boston: McGraw Hill Higher Education, 2010.

[6] Milter S A, Harley J P. Zoology. Seventh Edition. Boston: McGraw Hill Higher Education, 2007.

[7] Hartwell L H, Hood L, Goldberg M L, et al. Genetics: From Genes to Genomes. Fourth Edition. Boston: McGraw Hill Higher Education, 2008.

[8] Garrett R H, Grisham C M. Biochemistry. Fouth Edition. Belmont: Thomson Brooks/Cole, 2010.

[9] Molles M C. Ecology: Concepts and Applications. Fourth Edition. Boston: McGraw Hill Higher Education, 2006.

[10] Ratledge C, Kristiansen B. Basic biotechnology. Thrid Edition. Cambridge: Cambridge University Press, 2006.